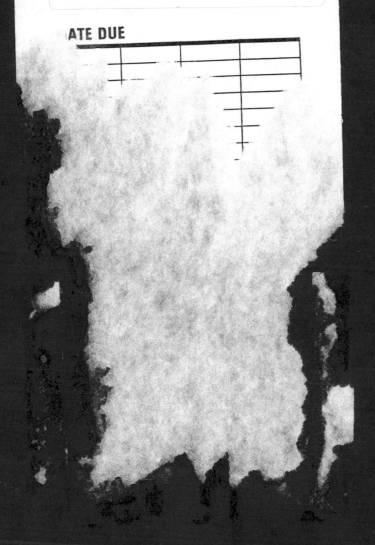

· *The Mapping of the American Southwest* ·

The Mapping
of the
American Southwest

EDITED BY

Dennis Reinhartz and Charles C. Colley

Texas A&M University Press

COLLEGE STATION

The paper used in this book meets the minimum requirements
of the American National Standard for Permanence
of Paper for Printed Library Materials, Z39, 48-1984.
Binding materials have been chosen for durability.

Library of Congress Cataloging-in-Publication Data

The Mapping of the American Southwest.

   Includes bibliographies.
      1. Southwestern States – Maps – Congresses.
   2. Cartography – Southwestern States – History –
   Congresses.   3. Southwestern States – Discovery and
   exploration – Congresses.   I. Reinhartz, Dennis.
   II. Colley, Charles C.
   GA408.5.S68M37   1987   912'.76   86-22992
   ISBN 0-89096-237-5

TO OUR GOOD FRIENDS

Jenkins and Virginia Garrett

Contents

List of Illustrations

· *xi* ·

Color Plates

Preface

*Words following words in long succession, however ably
selected those words may be, can never convey so distinct an
idea of the visible forms of the earth as the first glimpse
of a good map.*
— President, Royal Geographic Society, 1840

T he Mapping of the American Southwest," a symposium held
at the University of Texas at Arlington in the winter of 1983,
brought together cartographic experts from across the United
States to focus their attention upon one of the most dramatic but
often overlooked areas in the field of cartographic history—the American Southwest. It was significant that the meeting took place at the
Cartographic History Library of the University of Texas at Arlington, a center for the study of five centuries of exploration and mapping of the New World. Containing thousands of rare maps and atlases that feature the works of the great cartographers—including
Ptolemy, Coronelli, Delisle, Moll, and Arrowsmith—the library provided a grand setting for a symposium on origins of southwestern
cartography.

The symposium papers reproduced in this volume are varied in
scope and subject, but all relate significantly to the cartographic history of the southwestern frontier. The earliest period is covered in
"Spanish and French Mapping of the Gulf of Mexico in the Sixteenth
and Seventeenth Centuries" by David Buisseret, director of the Hermon Dunlap Smith Center for the History of Cartography at the
Newberry Library in Chicago. Buisseret espouses a broad and unique
synthesis of New World cartography; with the aid of a modern base
map and overlays, he shows where such noted cartographers and explorers as Pineda, La Salle, and Marquette erred in their concept of
geographical features of the Southwest and Gulf Coast regions. The
earliest maps were surprisingly accurate in some respects but highly

erroneous in others. For example, a mountain range was shown as an impenetrable barrier between the Eastern Seaboard and the lower South for many years. There was also great inconsistency in the location of the Mississippi River. From the sixteenth to the early eighteenth century, Spanish and French cartography went through four additional distinct phases, each leading to improved accuracy in delineation of rivers, mountains, and other geographic features.

Dennis Reinhartz, associate professor of history at the University of Texas at Arlington, takes up the study of eighteenth-century mapping in "Herman Moll, Geographer: An Early Eighteenth-Century European View of the American Southwest." Moll, a native of Germany, gained a reputation as a skilled cartographer in the service of English patrons. An extremely prolific, skilled engraver with an artistic flair, he attracted many imitators. In addition to single maps, he produced an abundance of atlases, globes, and charts of "surprising versatility." His influence upon the mapping of the American Southwest was great because he uniformly covered the Spanish, French, and British possessions of the New World, including the Gulf of Mexico. But Moll placed a good part of Texas within the boundaries of French Louisiana and continued to show California as an island long after most other cartographers accurately portrayed it as part of the North American mainland. Nonetheless, as this informative essay indicates, Moll's contributions to southwestern cartography were many, and his works added significantly to greater knowledge of the New World.

"United States Army Mapping in Texas, 1848–50," covers another facet of the subject in admirable detail; its author is Robert Sidney Martin, assistant director of libraries for special collections at the Louisiana State University in Baton Rouge and former director of the Cartographic History Library at the University of Texas at Arlington. Martin traces the roots of Texas cartography from the end of the Mexican War in 1848. At that time, the U.S. government's lack of accurate knowledge of the newly acquired lands of Texas led Secretary of War Jefferson Davis to send out several teams of army engineers to undertake surveys and to map routes for wagon roads westward. By 1849 their efforts had opened trails for use by California-bound travelers, and in 1850 they resulted in the publication of a large general map of Texas which greatly raised the standards of western cartography. It also set the stage for mapping the international boundary between the United States and Mexico in subsequent years.

Judith A. Tyner, professor of geography at California State University, also concentrates on nineteenth-century cartography in "Images of the Southwest in Nineteenth-Century American Atlases." Tyner notes that westward expansion soon led to American commercial cartography and American production of atlases. The early atlases often perpetuated old errors and myths, such as the "Great American Desert," which affected patterns of immigration and settlement. Slowly, however, the popular understanding of the American Southwest progressed from a "vast *terra incognita* . . . to one which closely resembled our own perceptions."

The careful research of these scholars illustrates that the cartographic history of the American Southwest is long, varied, and dramatic. The holdings of the Cartographic History Library at the University of Texas at Arlington also indicate that there are vast resources for study in this area. *The Mapping of the American Southwest* focuses attention on an area of cartographic history that deserves continued investigation.

For their continuing generous support of the University of Texas at Arlington and most especially its Special Collections, this volume is dedicated to Jenkins and Virginia Garrett. As a consequence of their efforts and the cooperation of the University of Texas at Arlington, all proceeds from this volume will go to setting up the Special Collections Publication Endowment to underwrite future publication activities of the Division of Special Collections of the Library of the University of Texas at Arlington. This volume is the first in what we hope will be a long and fruitful series.

Dennis Reinhartz
Charles C. Colley

Color Plates

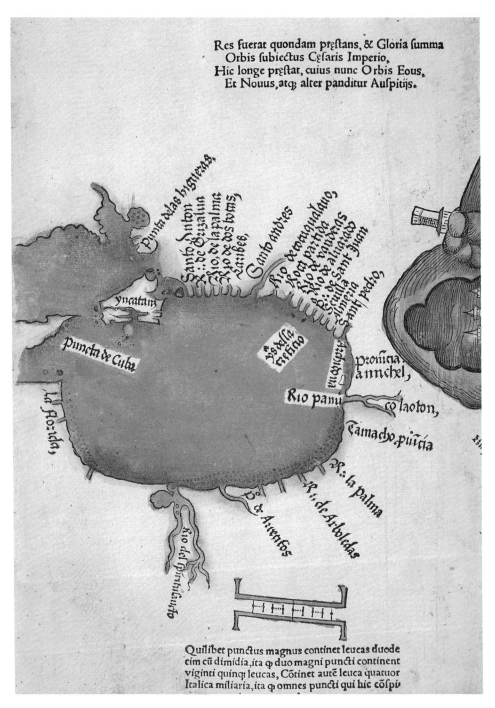

Res fuerat quondam prestans, & Gloria summa
Orbis subiectus Cesaris Imperio,
Hic longe prestat, cuius nunc Orbis Eous,
Et Nouus, atq3 alter panditur Auspitijs.

Punta delas higueras

Sant Anton

Ganto andres

yncatam

Puncta de Cuba

prouincia
annchel,

co laoton,

Rio panu

La florida

Çamacho puncia

Ri: la palma

Ri: de Arboledas

Po de Arrentos

Rio del spiritusancto

Quilibet punctus magnus continet leucas duode
cim cu dimidia, ita q3 duo magni puncti continent
viginti quinq3 leucas, Cotinet aute leuca quatuor
Italica miliaria, ita q3 omnes puncti qui hic cospi

Plate 1. Map of the Gulf of Mexico, printed in *Praeclara Ferdinandi Cortesii de nova maris oceani Hyspanis narratio* (Nürnberg, 1524). *Courtesy Newberry Library, Chicago*

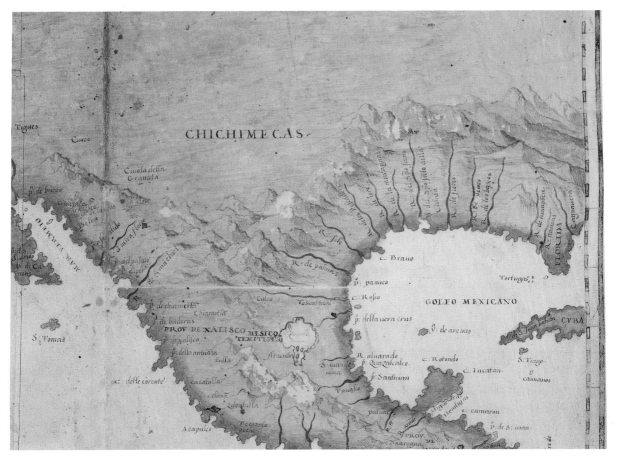

Plate 2. Detail from a map of the southern part of North America in a portolan atlas, probably by Battista Agnese, c. 1557. *Courtesy Newberry Library, Chicago*

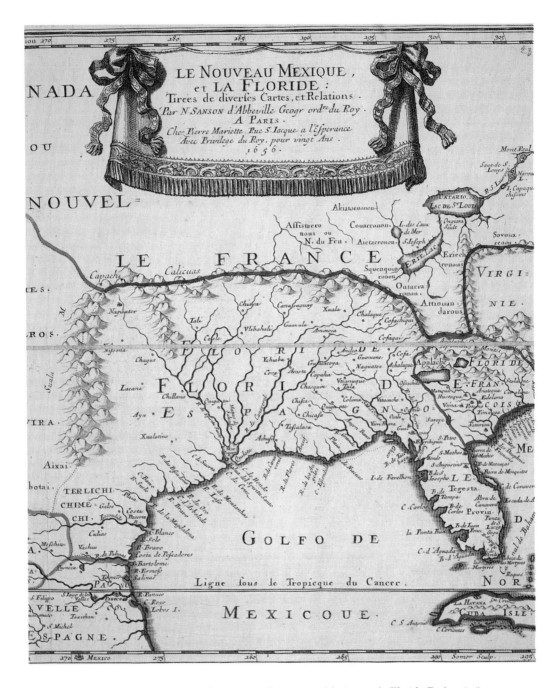

Plate 3. Detail from Nicolas Sanson, *Le nouveau Mexique et la Floride* (Paris, 1656). *Courtesy Newberry Library, Chicago*

Plate 4. "Carte du Mississippi et des rivieres qui s'y jettent . . . ," by Jacques Bureau. *Courtesy American Geographical Society's Collection at the University of Wisconsin–Milwaukee*

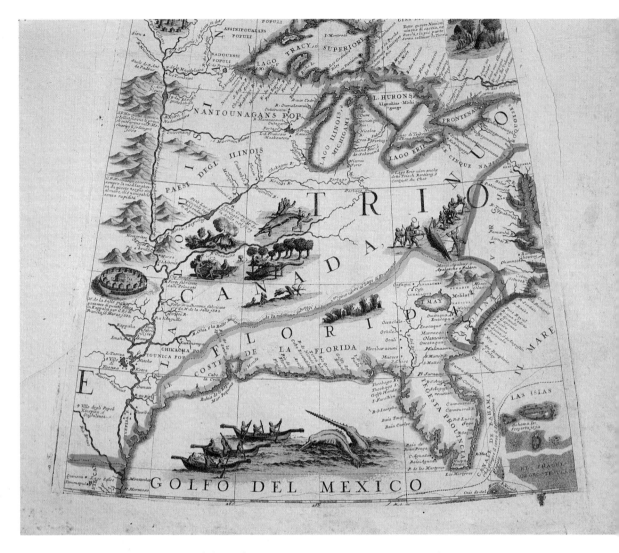

Plate 5. Detail from the map gore representing eastern North America, by Vincenzo Coronelli (Paris, 1688). *Courtesy Newberry Library, Chicago*

Plate 6. Detail from an anonymous map, *Carte nouvelle de l'Amerique angloise* (Amsterdam, c. 1700). *Courtesy Newberry Library, Chicago*

Plate 7. Detail from *Carte du Mexique et de la Floride*, by Guillaume Delisle (Paris, 1703). *Courtesy Newberry Library, Chicago*

· *The Mapping of the American Southwest* ·

Spanish and French Mapping
of the Gulf of Mexico in the Sixteenth
and Seventeenth Centuries

•

DAVID BUISSERET

The mapping of North America is a process which began with the Amerindians and quickened with the advent of the Europeans. It is a subject so vast that even its barest outlines could hardly be contained within one volume. That, no doubt, is why the most successful attempts at studying this complex process have been analyses of the maps of relatively small geographical areas. William Cumming, for instance, has examined the early maps of the southeast, of colonial Massachusetts, and of the Chesapeake Bay. Henry Wagner took for his area the northwest coast of America prior to 1800, and Carl Wheat investigated the Trans-Mississippi West.[1]

There have been similar works concerning the Gulf Coast. For the many volumes of his *Narrative and Critical History of America,* Justin Winsor compiled essays on regional cartography, including one on "the early cartography of the Gulf of Mexico and adjacent parts" and another on the "cartography of Louisiana and the Mississippi basin under the French domination."[2] Sixty years later, Jean Delanglez,

---

[1] William P. Cumming, *The Southeast in Early Maps* (Princeton: Princeton University Press, 1958); "The Colonial Charting of the Massachusetts Coast," in *Seafaring in Colonial Massachusetts,* vol. 52 of the Publications of the Colonial Society of Massachusetts (Boston: Colonial Society of Massachusetts, 1980); and "Early Maps of the Chesapeake Bay Area," in *Early Maryland in a Wider World,* ed. D. B. Quinn (Detroit: Wayne State University Press, 1982); Henry Wagner, *The Cartography of the North-West Coast of America to the Year 1800,* 2 vols. (Berkeley: University of California Press, 1937); Carl I. Wheat, *Mapping the Transmississippi West, 1540–1861,* 5 vols. (San Francisco: Institute of Historical Cartography, 1957–63).

[2] Justin Winsor, *Narrative and Critical History of America,* 8 vols. (Boston/New York: Houghton, Mifflin and Co., 1884–89), 2:217–30 and 5:79–86.

a Jesuit, brought much of his previous work together in *El Rio del Espíritu Santo*, a study of early cartography of the Gulf Coast.[3] James Bryan and Walter Hanak followed with *Texas in Maps*, and most recently Robert Sidney Martin and James C. Martin have surveyed four centuries of Texas and southwestern cartography.[4]

While these works contain information crucial to the study of Gulf cartography, none of them has the same objective as the present paper, which is to examine early maps of the Gulf not only in historical context but also as examples of five main phases of mapmaking. Each of these early phases produced a distinctive type of map which deserves further detailed study. Identifying these five main types required photocopying, assembling, and examining a great many maps. Some of these maps may be found only in rare publications or in unpublished form in their original repositories. Most, though, have been reproduced in one or more of four recent publications.[5] It should be added parenthetically that inexpensive methods of reproducing maps for study have made possible comparisons which Winsor, Parkman, and the other great pioneers could never have imagined.

Analyzing this sequence of maps required devising some means of comparing them with a modern base map. Many such methods have been suggested,[6] but all those which rely on comparison of points identified by their latitude and longitude had to be discarded, since

---

[3] Jean Delanglez, *El Rio del Espíritu Santo: An Essay on the Cartography of the Gulf Coast and the Adjacent Territory during the Sixteenth and Seventeenth Centuries* (New York: United States Catholic Historical Society, 1945).

[4] James Bryan and Walter Hanak, *Texas in Maps* (Austin: University of Texas Press, 1961); Robert Sidney Martin and James C. Martin, *Contours of Discovery: Printed Maps Delineating the Texas and Southwestern Chapters in the Cartographic History of North America, 1513–1930* (Austin: Texas State Historical Association, 1982).

[5] These publications are W. P. Cumming, R. A. Skelton, and D. B. Quinn, *The Discovery of North America* (London: Elek, 1971); W. P. Cumming, S. E. Hillier, D. B. Quinn, and G. Williams, *The Exploration of North America 1630–1776* (New York: G. P. Putnam's Sons, 1974); Adrian Johnson, *America Explored* (New York: Viking Press, 1974); and Seymour Schwartz and Ralph Ehrenberg, *The Mapping of America* (New York: H. N. Abrams, 1980).

[6] See, for instance, Elizabeth Clutton, "Some Seventeenth Century Images of Crete," *Imago Mundi* 34 (1982): 48–65; Harry Margary, "A Proposed Photograph Method of Assessing the Accuracy of Old Maps," *Imago Mundi* 29 (1977): 78–79; and William Ravenhill and Andrew Gilg, "The Accuracy of Early Maps? Towards a Computer Aided Method," *Cartographic Journal* 11 (1974): 48–52.

many early maps have no such coordinates. In the end, a system close to Harry Margary's "photographic method" was adopted. The base map was derived from the Albers's equal-area projection version of the Gulf. Then the various historical maps were projected onto this base map (using a Plan Variograph) in such a way that the Florida peninsula and the north-south section of the coast at the mouth of the Rio Grande made a "best fit." This method of comparison seems to be relatively accurate for east-west readings and, as will be seen, did not reveal any gross distortions in the north-south axis. For maps made when the coast was relatively well known, after about 1700, the base maps and the contemporary maps fit closely.

Type One, 1519–44

T he earliest map to show the Gulf of Mexico in any detail was the so-called Pineda chart, preserved at the Archivo General de Indias in Seville.[7] This map, drawn after the expedition of 1519 by Alonzo Alvarez de Pineda, is a marvelous testimony to the skill of Spanish cartographers at that time. Not only does it correctly show that there is no passage through the Gulf to the Pacific, but it also situates the Yucatán Peninsula and Cuba with remarkable accuracy. "La Florida" makes its first appearance, as does the mouth of the "rio del Espiritu Santo," roughly at the present mouth of the Mississippi.

Four or five years after the Pineda chart was drawn, the first printed map of the Gulf appeared. It was published at Nürnberg in 1524 with the second letter of Hernán Cortés, the *Praeclara Ferdinandi Cortesii de nova maris oceani Hyspanis narratio* (plate 1). In some respects, the map is less accurate than the Pineda chart; the Yucatán Peninsula, for instance, has become an island, and the outline of Florida is grosser. However, as figure 1-1 shows, the general conformity of the map to the actual coastline is remarkable. The "rio del Spirito Santo" is placed this time over a hundred miles east of the Mississippi Delta; with its twin forks it surely refers to the Tombigbee and Alabama rivers, emerging in Mobile Bay. Away to the west, another large forked river must be an indication of the Rio Grande. Typical of type one, this map

[7]Well reproduced in Schwartz and Ehrenberg, *Mapping*, p. 37.

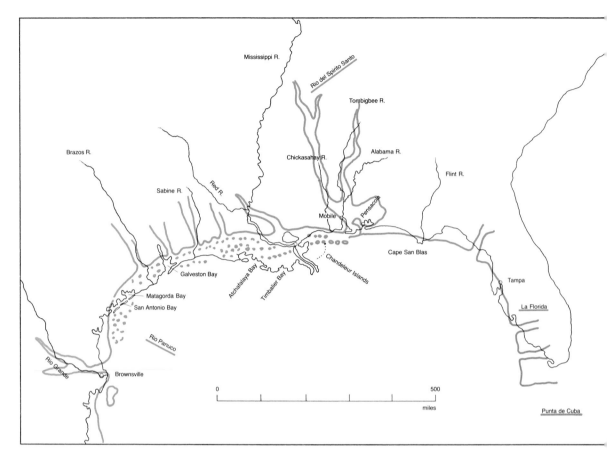

Fig. 1-1. Analysis of the 1524 map of the Gulf of Mexico by Hernán Cortés. *Courtesy Newberry Library, Chicago*

has a good general delineation of the coast, some indication of the main rivers, and very little information about the topography inland.

Maps of the Gulf Coast conformed to this type for the next twenty years. In 1525, for instance, the so-called Salviati map, dedicated to the Florentine family of that name,[8] followed the general delineation of the Cortés map but put the "Rio del Spiritu Santo" very close to the present mouth of the Mississippi. In the following year, the author of the Castiglioni map offered a very similar version of the Gulf Coast;[9] probably both maps were derived from the *padrón real,* the master map of the Spanish pilots based at Seville. In that same year, 1526, Juan Vespucci produced his superb planisphere.[10] Vespucci was one of the leading pilots at Seville, and his map is full of interest for the southeastern coast of North America. But it is rather disappointing for the Gulf Coast, where he has a multitude of ill-differentiated little rivers, and the large bay often identified with the Rio del Spirito Santo is shown very far to the west.

There survives a further cluster of maps from the late 1520s. Girolamo Verrazzano's world map of 1529 shows the Gulf Coast in some detail, but in a rather crude, squared-off style.[11] The superb world map of Diogo Ribero also shows the Gulf, in its characteristic type-one form, with the major bay about two hundred miles east of the Mississippi's mouth.[12] This map is in the Vatican Library, and a similar map may be found in the ducal library at Weimar.[13] The rather crude little woodcut "General map of the mainland and West Indies . . . ," published at Venice in 1534 and attributed to Ramusio, echoes the delineation of these maps.[14]

Nor is there any sign of new developments in the various products of the Dieppe School, first drawn in the mid-1530s. One of the earliest atlases of this school, preserved in the Royal Library at The Hague, has a superb chart of the Gulf of Mexico and Caribbean.[15]

[8] Cumming, Skelton, and Quinn, *Discovery,* p. 72.

[9] Ibid., p. 74.

[10] Ibid., pp. 86–87.

[11] Johnson, *America Explored,* p. 30.

[12] Cumming, Skelton, and Quinn, *Discovery,* pp. 106–7; Schwartz and Ehrenberg, *Mapping,* p. 40.

[13] On this cluster of maps, see Winsor, *Narrative and Critical History,* 2:221.

[14] Cumming, Skelton, and Quinn, *Discovery,* p. 71.

[15] Ibid., p. 58–59.

It marks an advance beyond most previous maps in no longer showing the Yucatán Peninsula as an island, but the Gulf Coast is rather
conventionally portrayed. There is a more convincing delineation on
the so-called Harleian Map of 1544, which has much in common with
the early Spanish maps, including a centered "R. de St Esprit" and
indications of the "R. de Palmas" and "R. de Panuco" roughly on the
site of the Rio Grande and the Pánuco rivers, respectively.[16]

However, even as the Dieppe cartographers were drawing their
superbly decorated maps, the Spanish captain Hernando de Soto was
leading an expedition which traversed much of the Gulf's coastal plain
and ought to have led to a much better understanding of its geography. Landing in Florida in May, 1539, de Soto reached the Mississippi
in May, 1541, and died the following year. The remnant of his fol-
lowers reached Pánuco in September, 1543, long after everybody had
given them up for dead.

Type Two, 1544–1680

Out of the experiences of de Soto's expedition came a new style
of map. Its prototype was the pen-and-ink sketch drawn about
1544 by the Spanish royal cartographer, Alonso de Santa Cruz, which
was very often reproduced.[17] The salient features of this map, ana-
lyzed in figure 1-2, are the internal rivers, delineated in far more detail than previously, and the ring of mountains to the north. Notice
that the "Baya de Spiritu Santo" is here shown only about one hundred miles east of the mouth of the Mississippi and that the "Rio de
Spirito Santo" is by far the largest of the rivers shown, eventually curv-
ing away very far to the east. Bryan and Hanak identify the other
named rivers as "R. de Palmas"—Rio Grande, "R. de la Madalena"—
Nueces River, "R. del Oro"—Trinity River, and "Rio de Montaños"—
Sabine River. They also identify eight of the clearly marked Indian
villages.[18]

[16]Ibid., pp. 150–51.

[17]Ibid., p. 121; Johnson, *America Explored*, p. 63; Schwartz and Ehrenberg, *Mapping*, p. 61.

[18]Bryan and Hanak, *Texas in Maps*, p. 5.

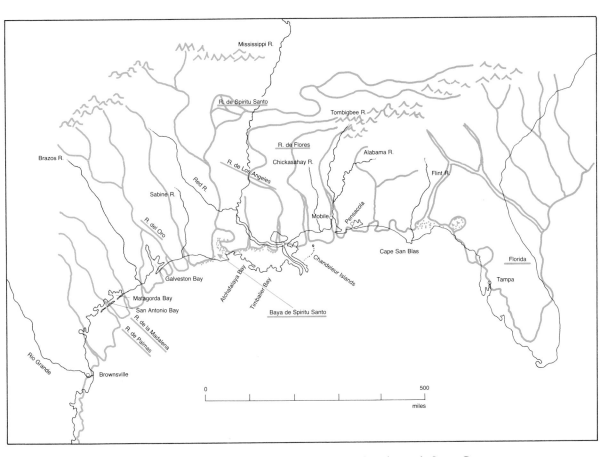

Fig. 1-2. Analysis of the 1544 map of the Gulf of Mexico by Alonso de Santa Cruz. *Courtesy Newberry Library, Chicago*

This type-two configuration, with a network of rivers and a mountainous barrier to the north, was to become the standard map for the next 130 years. (A few type-one maps continued to be produced,[19] but only until 1580.) One of the most extreme examples of the type-two configuration is the map thought to have been drawn about 1557 by the Genoese cartographer Battista Agnese (plate 2). Here the mountains positively prohibit any possibility of a great river entering the plain from the north. The "Rio de Palmas" and the "Rio de Spirito Santo" are shown close to the actual sites of the Rio Grande and the Mississippi. Much the same configuration is found on the printed map by Diego Gutiérrez, published in 1562,[20] and on Mercator's world map of 1569.[21] On the latter, the mountains bordering the Gulf Coast hem in the "Rio del Espirito Santo"; a great river is indeed shown emerging from the vicinity of the Great Lakes, but it then flows westward into the Pacific.

In the later sixteenth century, the most widely disseminated type-two map was that published in the *Additamentum III* to Abraham Ortelius's *Theatrum Orbis Terrarum*, 1584 edition. This map is ascribed to Gerónimo de Chaves, Spanish royal cosmographer who worked at Seville in the Casa de Contratación. It is less detailed in its account of the internal settlements than the Alonso de Santa Cruz map of 1544 and indeed marks the Indian towns quite differently.[22] The "Rio del Espiritu Santo" is shown entering the Gulf through a "Mar pequena" and flowing down from sources rising far away in the east. Could this be an echo of the Ohio before its junction with the Mississippi? Although in many respects inferior to the 1544 map, this one of 1584 was very influential, and not merely in successive editions of the *Theatrum Orbis Terrarum*. In 1597, for instance, Cornelius Wytfliet largely copied it in the "Florida et Apalche" map of his *Descriptionis Ptolemaicae augmentum*, the first printed atlas devoted to America.[23] Gabriel Tatton used its river pattern and internal names for his

---

[19]See, for instance, the map of 1580 illustrated in Cumming, Skelton, and Quinn, *Discovery*, p. 174.

[20]Johnson, *America Explored*, p. 164.

[21]The best reproduction seems to be in E. F. Jomard, *Les monuments de la géographie* (Paris: Duprat, 1842–62).

[22]Schwartz and Ehrenberg, *Mapping*, p. 73.

[23]Johnson, *America Explored*, p. 169.

Nova et rece (sic) terraum et regnorum Californiae . . . delinatio (1616), the map that looks like an advertisement for a variety of typographic styles.[24]

Toward the middle of the seventeenth century, French cartographers were coming into their own, and in their work we find the last two examples of the type-two map. In 1656 Nicolas Sanson published *Le nouveau Mexique et la Floride* (plate 3). Notice the impenetrable ring of mountains to the north, accentuated on this copy of the map by the green boundary line between "Nouvelle France" and "Floride espagnole." Notice, too, the agreeably symmetrical arrangement by which the "R. de Spirito Santo," the "R. de Canaveral," and several other rivers all flow into the "Mar pequeno." A very similar map, produced four years later by Pierre Duval, also had the centered "Mar pequeno" and the impenetrable mountains.[25] He, like Sanson, had a version of "Erie lac," which reminds us that by the 1660s the next phase in cartographic understanding of the Mississippi basin was about to begin.

By then the French settlements down the Saint Lawrence, established with such difficulty in the early seventeenth century, were beginning to push their priests and trappers out into the region of the Great Lakes and to map their complex hydrography. Between May and October of 1673, Louis Jolliet and Père Jacques Marquette undertook the voyage which confirmed that a great south-flowing river lay to the west of Lake Michigan, for they descended the Mississippi as far as its junction with the Arkansas River, where they were certain that they would soon reach the Gulf.

Type Three, 1673–84

On the way back to Montreal, Jolliet lost his maps and other documents when his canoe overturned. He therefore had to draw from memory the map that accompanied his report of the voyage. This map has been the object of much controversy. For many years a map in the John Carter Brown Library was thought to be

[24] Ibid., p. 168.
[25] Ibid., p. 174.

Jolliet's original; this seems not to be the case, and we have a choice among many copies of copies.[26] All, however, share one common feature, which indeed is the salient characteristic of a type-three map: they show the Mississippi descending straight down into the Gulf, in a north-south line from its confluence with the Arkansas River or even from the Missouri.

A relatively little-known but interesting manuscript example of this map is shown in plate 4. Preserved at the American Geographical Society collection in Milwaukee, it shows the Mississippi flowing straight down from the Great Lakes. It is known as the "Tonti" map, after La Salle's lieutenant, and seems to date from a somewhat later period. Notice that it is oriented north-south, which is understandable if it was conceived by or for somebody working down the Mississippi from the Great Lakes. Among the printed maps of this type, one of the most characteristic is found in the *Recueil de voyage de M. Thévenot* of 1681 (figure 1-3). The map is oriented westwards and shows the route that Jolliet and Marquette took from Lake Michigan (so named for the first time) to the lower Mississippi. Parts of it are rather gross, but it gives an excellent version of the Chicago portage between the Chicago and Desplaines rivers, site of the future Illinois and Michigan Canal.

The Thévenot map of 1681 presumptuously shows the Mississippi entering the Gulf. The *Carte de la Nouvelle France* . . . (figure 1-4), published two years later by Louis Hennepin, is more discreet. It follows the other type-three maps in its delineation of the upper and middle Mississippi but then indicates its course to the Gulf with only a dotted line. By 1683, though, that journey had in fact been accomplished by Cavelier de La Salle, and this new discovery soon made its way onto the maps, modifying them into type four.

Type Four, 1684–1702

La Salle seems to have passed on his cartographic information — or rather misinformation — to the young engineer, Jean-Baptiste Franquelin, then based at Québec. The map that Franquelin then

[26]See, for instance, ibid., p. 153, and Cumming et al., *Exploration*, p. 37.

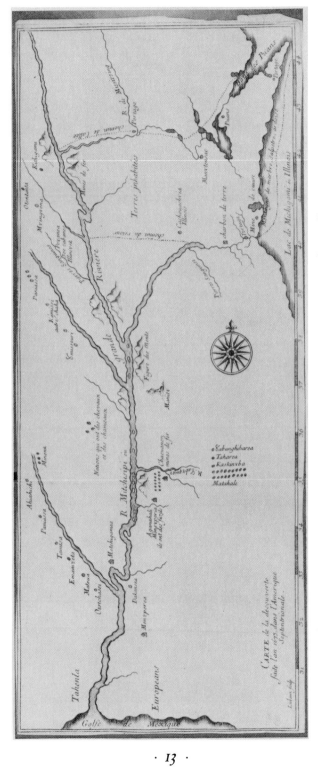

Fig. 1-3. Map from the Recueil de voyage de M. de Thévenot (Paris, 1681). Courtesy Newberry Library, Chicago

Fig. 1-4. Map from the *Description de la Louisiane,* by Louis Hennepin (Paris, 1683).
Courtesy Newberry Library, Chicago

drew, in 1684, appears to have been lost,[27] but it surely resembled his later versions, such as the 1687 *Carte de l'Amerique Septentrionale* preserved at the Service Historique de la Marine in Paris.[28] Their salient feature is that the mouth of the Mississippi is placed very far to the west, almost on the actual site of the Rio Grande. Some have maintained that La Salle deliberately misled Franquelin in this respect;[29] be that as it may, the map of 1684 inaugurated a new type.[30]

Many of these type-four maps appear to have been lost and are known to us only in the shape of rather dubious tracings. But one particularly fine example has survived; it is the map gore representing eastern North America produced by Vincenzo Coronelli in 1688 (plate 5). Here the Mississippi runs into the sea just north of the mouth of the Rio Bravo. The "R. Ohio" is therefore obliged to take its (dashed) course very far to the south of where it actually runs, and the Illinois River becomes extremely long. These distortions aside, however, Coronelli's was a fine map, with a particularly good delineation of the Great Lakes, several of which still bear their earliest names.

Nine years later Louis Hennepin published his *Carte d'un tres grand pays entre le Nouveau Mexique et la mer glaciale.*[31] Although the Great Lakes and Hudson's Bay were less well delineated than on the Coronelli map, Hennepin's version of the Ohio River and of the eastern seaboard was better. Perhaps the apotheosis of this type-four map was by the anonymous craftsman of the *Carte nouvelle de l'Amerique angloise,* published at Amsterdam about 1700 (plate 6). Here the "Mitchisipi" and the "Rio Grande" are one, emerging from the corner that the east-west coast forms as it meets the north-south coast coming up from Mexico.[32]

---

[27] There is a copy of it in Winsor, *Narrative and Critical History,* 4:228.

[28] Cumming et al., *Exploration,* p. 153.

[29] See most recently Louis De Vorsey, "The Impact of the La Salle Expedition of 1682 on European Cartography," in *La Salle and His Legacy,* ed. Patricia Galloway (Jackson: University Press of Mississippi, 1982).

[30] As usual, some of the old type continued to be produced; see, for instance, the "Locke map" of Carolina, reproduced in Cumming et al., *Exploration,* p. 88.

[31] Johnson, *America Explored,* p. 157.

[32] For a remarkable example of a type published long after it had been discredited, see the "type four" map of J. F. Bernard, *Le cours du fleuve Mississippi selon les relations les plus modernes* (Amsterdam, 1737), reproduced in *Naissance de la Louisiane* (Paris: Ministère de la Culture, 1982), p. 41.

Type Five, 1700–

B y the late 1690s, the lower Mississippi was becoming the center of a power struggle between the French and the Spaniards, in the course of which its cartography was bound to become better known. La Salle had returned there in 1684, leading an expedition which, coming from the sea this time, disastrously failed either to find the mouth of the Mississippi or to establish an enduring colony. The Spaniards, having heard of the venture but not of its failure, between 1685 and 1688 sent out several expeditions hunting for the colony. Then in 1698, to forestall further French attempts to establish themselves at the mouth of the Mississippi, the Spaniards occupied Pensacola from the east. In the spring of the following year, Iberville founded Biloxi.

All this activity produced a good deal of information about the geography of the Mississippi Valley. Much of it was relayed back to France, where in 1700 the celebrated cartographer Guillaume Delisle published in the *Journal des savants* an open letter,[33] explaining to "M. de Cassini" his latest views on the location of the mouth of the Mississippi. There had not as yet been, he wrote, any astronomical observations for this area, which would give certainty in the matter. However, relying on information from the various Spanish expeditions, beginning with Pineda, on maps by La Salle, and on conversations with the survivors of that expedition, Delisle claimed to have located the mouth in a place recently confirmed by the Iberville expedition.

Delisle summarized his new view of the Mississippi Delta on his *Carte du Mexique et de la Floride* of 1703 (plate 7), which is the prototype of our fifth and final map type. This map conformed very closely to the actual outline of the coast. With Tampa fixed in the east and the mouth of the Rio Grande in the west, we recognize all the landmarks in their correct position: Matagorda Bay, the mouth itself, Mobile, Pensacola, and even many of the internal rivers.[34]

---

[33] See *Journal des savants* (1700): 211–18 and Bibliothèque Nationale (Paris), manuscrits français 9097, fo. 97.

[34] About the same time, Franquelin also drew his superb "Carte de la nouvelle France," preserved at the Bibliothèque Nationale, Cartes et Plans GEDD 2987 (8536), and reproduced in *Naissance de la Louisiane*, p. 31.

Much remained to be done and would be the work of the next few decades: precisely delineating the channel of the Mississippi, accurately tracing the rivers, marking the site of Indian settlements, and so forth. The story of this later work is a complex and interesting one, and it remains untold. But for our purposes we need go no further; the attempt to portray the northern Gulf, so brilliantly begun by Pineda in 1519, reached its culmination nearly two centuries later in the first generally reliable map.

Herman Moll, Geographer:
An Early Eighteenth-Century European View of the American Southwest

•

DENNIS REINHARTZ

Introduction: The Life and Work of Herman Moll

In the late seventeenth and early eighteenth centuries, the growing affluence, literacy, and leisure of the rising Western European middle classes brought about increased interest in books and maps. Influenced by the prevailing quasi-scientific worldview of the later Enlightenment, the reading public demanded greater knowledge of distant and exotic places through travel literature, often illustrated, of real and fictional voyages and adventures. By 1750 hundreds of such works were readily available to tell cultured Europeans, sometimes with questionable accuracy, about new discoveries, different lands, and strange peoples from around the world.[1]

In Great Britain cartographers like Herman Moll helped to shape the popular images of Africa, Asia, and the Americas. Moll probably

I gratefully acknowledge the research fellowship from the Hermon Dunlap Smith Center for the History of Cartography at the Newberry Library, which allowed me to carry out the bulk of the research for this project in its excellent collections during the summer of 1982. I also am deeply indebted to the director and staff of the Cartographic History Library of the University of Texas at Arlington for their continuing support of my research.

[1]In addition to the sources cited in this chapter and the cartobibliography of Herman Moll provided in the appendix to this book, relevant works include: Percy G. Adams, *Travelers and Travel Liars, 1660–1800* (New York: Dover Publications, 1980); Thomas Chubb, *The Printed Maps in the Atlases of Great Britain and Ireland: A Bibliography, 1579–1870* (London: Homeland Association, 1927); John Green, *The Construction of Maps and Globes* (London: T. Horne et al., 1717); R. V. Tooley, *Maps and Map Makers* (New York: Crown Publishers, 1978).

was born in Germany in 1654. Little is known of his life until 1678, when he settled in London.[2] He first gained notice as a fine engraver working for Moses Pitt, Greenville Collins, John Adair, Seller & Price, and others; his excellence in engraving earned him a substantial reputation and many imitators. Later his reputation also grew as a geographer and bookseller. He established himself in Vanley's Court in Blackfriars and finally in Devereux Court in the Strand and often employed George Vertue as his engraver. In 1708 Moll founded *Atlas Geographagus,* a unique monthly magazine which ran until 1717, when it was republished in five volumes. It was soon imitated by others, including James Knapton and Daniel Defoe. Moll was associated with Sir William Stukeley and his intellectual circle and came to be seen by many in Britain as "a great mapmaker" and "the foremost geographer of his time."[3] Among Moll's intimates were scientists like Robert Hooke, the discoverer and namer of the cell, the buccaneers William Dampier and Woodes Rogers, literati such as Jonathan Swift and Defoe, and several important publishers, Knapton among them. Moll died on September 22, 1732.

By the time Moll began working, copper engraving had been established in Europe for over a hundred years. He was a prolific cartographer, and over a career of more than a half-century his maps, charts, atlases, and globes exhibited a surprising diversity. The earliest known maps by Moll are probably "America" and "Europe," which appeared in Sir Jonas Moore's *A New Systeme of Mathematiks Containing . . . a New Geography . . .* , published in London in 1681. An early Moll chart was in Collins's *Coasting Pilot* of 1686. He did nine charts for Adair in 1688 and six more for Collins in 1689.[4] But his first major work, *A System of Geography . . .* , was published in London in 1701. It was a global geography with a full complement of maps. This was

---

[2] William Stukeley, *The Family Memoirs of the Rev. William Stukeley, M.D.,* 3 vols. (London: Surtus Society, 1882–87), 1:134; Sarah Tyacke, *London Map-Sellers, 1660–1720* (Tring, Herefordshire: Map Collector Publications, 1978), p. 123; J. N. L. Baker, "The Earliest Maps of H. Moll," *Imago Mundi* 2 (1937): 16. Wilhelm Bonacker asserts that Moll was from Bremen, but offers no proof. See *Kartenmacher aller Lander und Zeiten* [Cartographers of All Lands and Times] (Stuttgart: Anton Hiersemann, 1966), p. 169.

[3] Willard Hallam Bonner, *Captain William Dampier: Buccaneer-Author* (Stanford, Calif.: Stanford University Press, 1934), p. 29.

[4] Baker, "Earliest Maps," p. 16.

followed by *The Complete Geographer . . .* and *Atlas Manuale . . .* in 1709, *A View of the Coasts Countries and Islands within the Limits of the South Sea Company . . .* in 1711, *Atlas Geographagus . . .* and *The World Described . . .* in 1717, *Atlas Minor . . .* in 1727, and numerous other maps and atlases in multiple editions which were local, regional, or global in scope. Most maps were printed separately first, usually in two sheets, and then bound and published later as atlases in joint undertakings between Moll and several other London publishers.[5] Depending on the number of additions, erasures, and strikes, the life of the average copperplate in the eighteenth century could be considerable.[6] Some of Moll's were used for more than thirty years.[7]

In addition to the more usual contemporary maps, charts, and atlases, Moll also produced several other interesting items. In 1703 he created his first two globes, one terrestrial and one celestial. His *Thirty New and Accurate Maps of the Geography of the Ancients . . .*, published in London in 1726, was actually a historical atlas to accompany the Greek and Roman classics. He also did numerous plans of cities, including London and Gibralter. Yet some of Moll's most unique works appear in books by others. "A View of the General Trade-Winds, Monsoons or Shifting-Winds & Coasting-Winds through ye World, Variations &c. . . ." in *Modern History . . .* (1725), by Thomas Salmon, and "A View of General & Coasting Trade-Winds in the Great South Ocean . . ." in *Collection of Voyages . . .* (1729), by Dampier, "England's greatest buccaneer" (who was responsible for marooning Alexander Selkirk, the model for Robinson Crusoe), are good early examples of thematic mapping.

Moll's career ended just as John Harrison developed his accurate chronometer to permit precise measurement of longitude—a function of time—and better topographic mapping. Moll produced no real examples of topographic maps, and his work does not reflect the new ideas on longitude: in fact, he thought topography dealt with "trivialities." On the other hand, Moll also did a world map for the fourth

---

[5] Henry N. Stevens, *The World Described in Thirty Large Two-Sheet Maps by Herman Moll, Geographer* (London: Henry Stevens, Son & Stiles, 1952), p. ii.

[6] For an excellent discussion of copperplate map printing, see Coolie Verner, "Copperplate Printing," in *Five Centuries of Map Printing*, ed. David Woodward (Chicago: University of Chicago Press, 1975), pp. 51–75.

[7] Stevens, *World Described*, pp. ii–vi.

edition of Defoe's *Robinson Crusoe*,[8] and the fictional maps for the first edition of *Gulliver's Travels* were "traced" directly from real ones by Moll.[9] He is mentioned by Gulliver in the book, along with Dampier—the only two contemporaries so honored by Swift.[10] While Moll probably lent his maps willingly to his literary friends, there were no copyright laws in England until 1734, and his maps were pirated regularly. Thus the expanse of Moll's mapping exposed the real and fictional earth and gave impetus to further charting of the unknown.

Not only was Moll's cartography extensive, but it also was distinctive. His style, lettering, and selection and treatment of subject matter contributed to truly unique and revealing representations (fig. 2-1). During Moll's time geography was usually defined as cartography, the locating of places. But Moll felt that "geography alone is dry and jejune and makes but a small impression on the memory"; hence he added natural history and descriptions to his works.[11] In his recent book, *Early Maps*, Tony Campbell writes: "No other maps tell us as much about their author as those of Herman Moll. . . . His early work is neat and restrained. It is only when he started to publish a series of two-sheet maps [*The World Described* . . .] . . . that his personality begins to shine through; for Moll could not resist airing his theories and prejudices in long notes scattered over his maps."[12] He even offered advice of a nongeographic nature. Nowhere is this more true than on his maps depicting the American Southwest.

Moll's numerous maps showing the American Southwest can be categorized into nine groups:

[8] Herman Moll, "A Map of the World, on wch [*sic*] is Delineated the Voyages of Robinson Crusoe," in Daniel Defoe, *The Life and Strange Surprising Adventures of Robinson Crusoe, of York, Mariner* . . . , 4th ed. (London: W. Taylor, 1719), pp. 80–81.

[9] Frederick Bracher, "The Maps in Gulliver's Travels," *Huntington Library Quarterly* 8 (1944): 60–64. See also Dennis Reinhartz, "The Cartographer and the Literati," *Mapline* 28 (December 1982): 3–4.

[10] See Jonathan Swift, *Travels into Several Remote Nations of the World, by Lemuel Gulliver, First a Surgeon, and Then a Captain of Several Ships*, 2 vols. (London: Benj. Motte, 1726).

[11] P. J. Marshall and Glyndwr Williams, *The Great Map of Mankind: Perception of New Worlds in the Age of Enlightenment* (Cambridge, Mass.: Harvard University Press, 1982), pp. 47–48.

[12] Tony Campbell, *Early Maps* (New York: Abbeville Press, 1981), p. 37.

Fig. 2-1. Beaver inset from "A New and Exact Map of the Dominions of the King of Great Britain on ye Continent of North America . . . ," in *The World Described . . .* (1715). *Courtesy Mr. and Mrs. Jenkins Garrett, Fort Worth*

1. World maps in "planisphere" (double hemisphere) or cylindrical, modified Mercator projections. Thematic tradewinds and ocean currents variations of both projections are included in this category

2. South Sea Company maps of the South Atlantic and South Pacific area, showing the southern part of North America

3. Sea Coast of Europe, Africa, and America charts

4. America maps of the western hemisphere

5. North America maps delineating the Spanish, French, and British empires

6. Western North America maps of the Spanish and French possessions

7. New France maps of French North America

8. New Spain maps of Spanish North America

9. Gulf of Mexico maps of the Western Gulf and Central America, the Northern Gulf and Florida, or the West Indies

The maps in categories 1–4 show little detail, especially of the interior, but they do record geographical and political boundaries. Moll characteristically showed California as an island and included most of Texas in Louisiana or Florida. Category 5 maps show more coastal features and some interior detail. But because of their greater specifics and notes, the maps in categories 6–9 are the most significant for understanding Moll's view of the Southwest.

California as an Island

One of the most distinctive features of Moll's Southwest is the portrayal of California as an island. Although this was common among the cartographers of the time, few of Moll's contemporaries continued to show California in this manner after its insularity had been disproved. Yet with two exceptions Moll carried out this portrayal until his death, and it was posthumously perpetuated until the 1780s through the reissue of his *The World Described* . . . and *Atlas Minor*. . . .

The California misconception was born with early Spanish explorers like Juan de la Fuca and Martin d'Aguilar and was initially shown on a 1620 map by Carmelite father Antonio Ascension. More

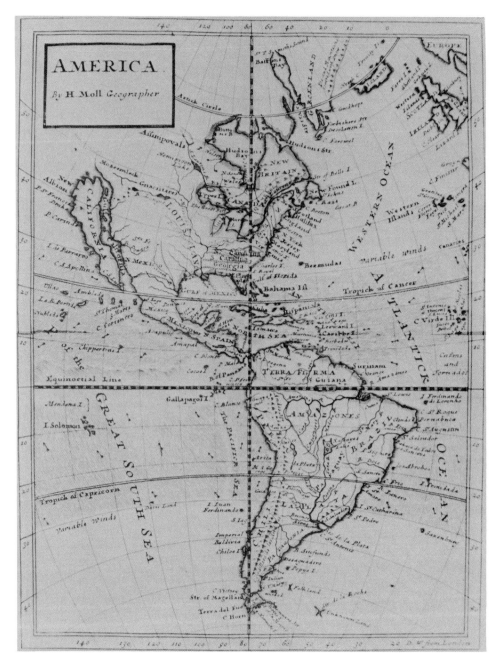

Fig. 2-2. "America," in *Atlas Manuale . . .* (1713). *From the private collection of Dennis and Judy Reinhartz*

than seventy-five years later a Jesuit, Father Eusebio Kino, was the first European to journey overland from California; a map he drew in 1698, published in 1705, showed California as part of the mainland. Father Kino's view at first did not find general acceptance, especially with Moll, who commented, "Why, I have had in my office mariners, who have sailed around it."[13] But in the multivolume 1717 edition of *Atlas Geographagus . . .* , Moll too questioned California's insularity:

> The first Discovery is ascribed to Fernando Cortez, in 1535. It was long doubted whether it was an Island or Part of the Continent, and does not seem yet to be fully determined. The first probability of its being an Island, proceeded from a Draught of it taken on Board a Spanish Ship by a Dutch Man: But Captain Rogers who was here in 1709, still doubts it. Some Spaniards told him, that several of their Countrymen had sailed as far up betwixt California and the Main as Lat. 42. where meeting with shoal water and abundance of Islands, they durst venture no further. Therefore he thinks it may probably join thereabouts to the Continent, because shoal Water and Islands is a general Sign of being near some Main Land. Dampier says, that the Spaniards in some of their late Draughts join California to the Main.[14]

Yet in the same volumes, the two maps showing California in its entirety portray it as an island. (The matter was officially settled in 1747, when Ferdinand VII of Spain decreed that California was not an island.)

Only rare cartographers were also explorers who based their maps on their own observations. Particularly in the eighteenth century, most were collators whose maps summarized current information supplied by others—explorers, travelers, fellow cartographers, and the like. Moll once explained: "As the Knowledge of Foreign Countries is a Science that no Man of either Learning or Business can excusably be without, so there is no certain way of attaining it but by consulting the Travellers that have been upon the Spot."[15] The general shape

---

[13] R. V. Tooley, *California as an Island, Map Collector's Series No. 8* (London: Map Collector's Circle, 1964), pp. 2–3.

[14] Herman Moll, *Atlas Geographagus; or, A Compleat System of Geography (Ancient and Modern) for America*, vol. 5 (London: Eliz. Nutt, 1717), pp. 675–76.

[15] Herman Moll, *The Complete Geographer; or, The Chorography and Topography of All the Known Parts of the Earth* (London: 1709), "Advertisement" page at the beginning.

Fig. 2-3. "The Isle of California . . . ," in *Atlas Manuale . . .* (1713). *From the private collection of Dennis and Judy Reinhartz*

of Moll's island of California probably is based on the second representation (1656) of the French cartographer Nicolas Sanson, also known as d'Abbeville, who is cited as a source in *Atlas Geographagus*... and other works.[16] Moll provided little of California's coastal detail and less of the interior, although on a 1720 map of North America he put an inset of New Albion harbor according to Sir Francis Drake.[17]

Although Moll's maps generally show California as an island, there are two interesting exceptions. One occurs in an undated edition of the *Atlas Minor*..., a copy of which is in the Edward E. Ayer Collection of the Newberry Library. On the first map of the volume, a world map, California is a peninsula, but on later maps of America, North America, New France, and New Spain in the same volume, California remains an island.[18] It may be that the plate of the world map in question was updated, perhaps by someone other than Moll, while the rest of the maps were not, for a posthumous reissuing of the *Atlas Minor*... after 1763.

The second exception is in Emanuel Bowen's *A Complete System of Geography*..., published in London in 1747. The maps in this work were from or were based on those in the 1723 edition of Moll's *The Complete Geographer*..., and California is not an island on any of them. Clearly, they were corrected.[19] In fact, in the text Bowen addressed the question of California and concluded with absolute surety that it was not an island:

> It was a Matter of Doubt for a long Time, whether it was an Island or a Peninsula. It used to be laid down in the best Maps as an Island, with a pretty wide Sea betwixt it, and the Continent of New Mexico; but in the latest Maps it is laid down as a Peninsula, which it

[16] Moll, *Atlas Geographagus*, pp. 675–81. See also Tooley, *California as an Island*, pp. 23–24.

[17] Herman Moll, "To the Right Honourable John Lord Somers Baron of Evesham in ye County of Worcester President of Her Majestys most Honourable Privy Council &c. This Map of North America According to ye Newest and Most Exact Observations is most Humbly Dedicated by your Lordships Most Humble Servant Herman Moll Geographer . . . ," *The World Described; or, A New and Correct Sett of Maps*... (London: 1720–54, and Dublin: 1730–41).

[18] Herman Moll, *Atlas Minor*... (London: n.d.), pp. 1, 47–48, 50, 54.

[19] See Emanuel Bowen, *A Complete System of Geography*..., 2 vols. (London: Innys, 1747), 1:vii–ix, 2:520–21, 598–99, 620–21, and 740–41. See also Herman Moll, *The Complete Geographer* (London: 1723).

really is, though it did not appear to be a Certainty, till it was dis-
covered to be such by Father Caino (or Kino, as he spelt it in our
Map) a German Jesuit, who landed in California from the Island of
Sumatra, and passed into New Mexico without crossing any other
Water than Rio Azul, or the Blue River, about North Latitude 35.[20]

New Mexico and New Spain/Mexico

California certainly is not the only pecularity of Moll's Southwest. His interior studies, especially those of Texas, are even
more distinctive. In Moll's work there is a wealth of detail and notation, but no one map or set of maps encompasses it all, nor is the
information organized or presented progressively over a series of maps
of the same area. Maps of the same set or volume also were not necessarily created at the same time for one purpose, and updatings oc-
curred individually and irregularly.

As in the case of California, Moll's sources were numerous and
diverse. From citations in *Atlas Geographagus* . . . and elsewhere, Moll's
major sources for the southwestern interior seem to have been Sanson and his sons, the French cartographer Alexis Hubert Jalliot, the
Dutch cartographer Jan Luyts, and Dampier.[21] Moll was the engraver
for Baron Lahonton's American map of 1703, which perpetuated the
myth of a great salt lake near California.[22] Subsequently, this and
other fantastic material was carried over to Moll's own maps. For example, "Mozeemleck," "Gnactsitares," and the "R. Longue" flowing
eastward into the Mississippi River appeared on Moll's "America . . ."
map in the *Atlas Minor* . . . as late as 1736.[23]

On various American maps Moll indicates individual details of
the interior of New Spain/New Mexico (sometimes generally labeled
"Parts Unknown"). On a 1719 North American map from *The World
Described* . . . , a "San Antonio" is placed near "Socorro," and he men-

---

[20]Bowen, *Complete System*, 2:619.

[21]Moll, *Atlas Geographagus*, pp. 454, 582–681.

[22]Baron de Lahonton, *Nouveaux Voyages de Mr. de Baron de Lahonton dans
L'Amerique Septentrionale* . . . (The Hague: 1703).

[23]Herman Moll, *Atlas Minor* . . . (London: 1736). See also John Lieghly, *California as an Island* (San Francisco: Book Club of California, 1970), p. 147.

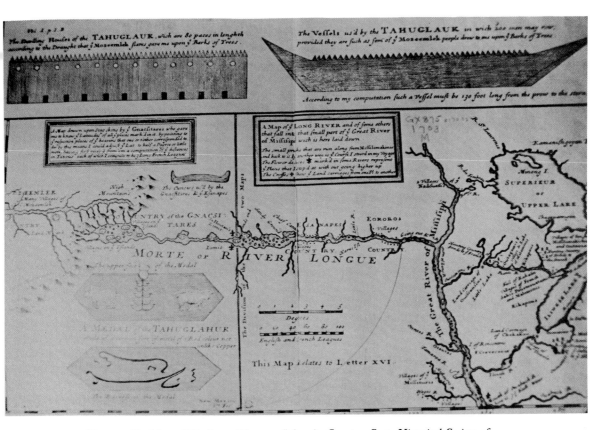

Fig. 2-4. "A Map of Ye Long River . . ." (1703). *Courtesy State Historical Society of Wisconsin, Madison*

tions the puebloes "Tzia" (Zia), "Taos," "S. Clara," "Acoma," "Zuni," and Coronado's legendary "Quivira" (probably in present-day Kansas), among others.[24] In the 1717 *Atlas Geographagus . . .* the fabulous "Cibola" itself is cited.[25] "Santa Fe" also is designated on most maps, as is "El Passo," but the latter refers to the actual pass through the mountains and not the more modern city. On a New France map in *Atlas Geographagus . . .* there is a reference to "Sandia."[26] Whether it marks the so-named mountains just east of present-day Albuquerque or the melons that the Indians raised in the upper Rio Grande Valley, near the site of the future New Mexican city, is unknown. Numerous Indian tribes are also designated. On a chart in *Atlas Geographagus . . . ,* for example, the "Apaches" and their cousins, the "Navahoes," are listed in "West Florida" (New Mexico) near the "North River" (Rio Grande).[27]

Texas

Although Moll generally presented Texas in much the same manner as the rest of the Southwest, these maps have generated controversy. The dispute centers on the eastern boundary of Texas, a segment of the border between Spanish and French North America, in the first half of the eighteenth century.

As late as 1736, Moll without exception put a large part of Spanish Texas in French Louisiana. While the French officially claimed only the territory east of the lower Red River ("R. Rouge"), Moll put the line from the Pecos River ("R. Salado") south to the Rio Grande ("R. del Norte") and on to the Gulf of Mexico. This distortion undoubtedly is based on that of Vincenzo Maria Coronelli, "the Cosmographer of the Venetian Republic," who also worked for Louis XIV of France in the 1680s.[28] Coronelli's distortion in turn was based on misinformation from La Salle. But Moll, often having to rely on French cartographers because Spanish maps did not exist or were un-

---

[24] Moll, ". . . North America . . . ," *The World Described; New Maps*, p. 7.
[25] Moll, *Atlas Geographagus*, pp. 671–74.
[26] Ibid.
[27] Ibid., p. 454.
[28] Tooley, *California as an Island*, pp. 2–3.

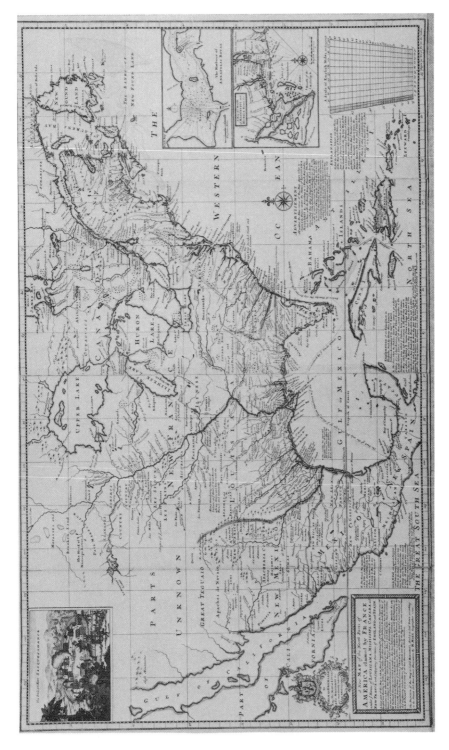

Fig. 2-5. "A New Map of the Parts of North America claimed by France . . . ," in *The World Described . . .* (1726). *Courtesy Cartographic History Library, University of Texas at Arlington*

available, updated and enhanced the errors in mapping Spanish America. La Salle's "French Fort" was indicated on "St. Bernard or St. Louis Bay" (Galveston Bay), as was the Spanish presence in Louisiana on a 1715 map of the West Indies.[29] On an earlier map in *A System of Geography . . .* , the interior of French Texas near the Rio Grande contains the notation that "About this Riv. The People [French and/or Indians?] are continually in Wars with the Spanyards."[30] In the same volume consistency again gives way to confusion when the "R. Bravo" (Rio Grande) is listed as a "principal River of Spanish Florida."[31] On a map of New France in *The World Described . . .* , Moll also traced the routes of the French explorer Louis Juchereau de St. Denis from Natchitoches in Louisiana across Texas to the "R. del Norte" in 1713–16.[32]

The St. Denis information probably came from Guillaume Delisle's famous *Carte de la Louisiane et du Cours du Mississipi . . .* of 1718, on which the name Texas ("Tejas") also first appeared. This brings up an interesting problem. Delisle, the premier French cartographer of his day, produced some of the most accurate maps of North America. But his delineations of often vague boundaries between the French, British, and Spanish empires on this continent created quite a furor in this era of early nationalism. The French royal cartographer's 1718 map not only incorporated part of the Spanish Southwest (Texas) into French Louisiana in the west but also claimed the Carolinas in the east for France. Moll and other British cartographers were willing to accept French claims to Texas, though contemporary Spanish cartographers rejected such maps, but the British took exception to French claims to the Carolinas. They responded with their own maps, including Moll's British North America map of 1715 and his North America map of 1720, among others.[33] The 1720 map specifically attacked

[29] Herman Moll, "A Map of the West Indies or the Islands of America in the North Sea . . . ," *The World Described; New Maps.*

[30] Herman Moll, "The Isle of California. New Mexico. Louisiane. The River Misisipi. and the Lakes of Canada . . . ," *A System of Geography; or, A New & Accurate Description of the Earth in All Its Empires, Kingdoms and States* (London: T. Childe, 1701), p. 152.

[31] Ibid., p. 155.

[32] Herman Moll, "A New Map of the North Parts of America claimed by France under ye names of Louisiana, Mississippi, Canada and New France with ye Adjoining Territories of England and Spain . . . ," *The World Described; New Maps.*

[33] Herman Moll, "A New and Exact Map of the Dominions of the King of

Delisle's 1718 map and the French claims to British North America, yet it left the Spanish-French boundary unchanged. In fact, Delisle's maps were often appended to later editions of Moll's *The World Described* . . . , undoubtedly so that the purchasers might compare for themselves. As Campbell indicates, "the seeds of the Anglo-French rivalry that were to lead to the French and Indian War of 1754 had clearly already been sown."[34] Spanish cartographers also were forced to produce new maps to refute French claims; sometimes they even had to publish cartographic information about the Southwest which Spain had hitherto closely guarded as state secrets.

Commenting on Moll's Texas-Louisiana distortion, James P. Bryan and Walter Hanak assert in *Texas in Maps* that, "Aside from this outstanding representation, Moll contributed little to Texas cartography."[35] Their judgment is unduly harsh, for Moll's mapping of Texas and northern Mexico is both informative and appealing. He was best at coastal geography, depicting with some accuracy the coastal features, barrier islands (e.g., Padre Island), and identified rivers emptying into the Gulf of Mexico. The rivers often continue deep into the interior, where there is less detail, but Moll does indicate various Indian tribes, along with the "Silver Mines of Caouila" in northern Mexico.[36] One map also shows the mysterious "Vil. of White Spanards" and "Vil. of Black Spanards" in Central Texas (Louisiana).[37]

But most intriguing are Moll's notations. For example, he mentions several times the Spanish cattle gone wild—the famous Texas longhorns of later years—by noting "Country full of Beeves" or "This Country has vast and Beautiful Plains, all level and full of Greens, which afford Pasture to an infinite number of Beeves and other Creatures" in East Texas near the "R. Salado." Nearby also is noted, "Many Nations [of Indians] on ye heads of this Branches [of several rivers] who use Horses and Trade with the French and Spanjards." Off the

---

Great Britain on ye Continent of North America . . . ," and ". . . North America . . . ," *The World Described; New Maps.*

[34]Seymour I. Schwartz and Ralph E. Ehrenberg, *The Mapping of America* (New York: Harry N. Abrams, 1980), pp. 35, 138–41, 146; Campbell, *Early Maps*, p. 37.

[35]James P. Bryan and Walter Hanak, *Texas in Maps* (Austin: University of Texas Press, 1961), p. 8.

[36]Moll, "A Map of the West Indies . . . ," *The World Described; New Maps.*

[37]Moll, ". . . North America . . . ," *The World Described; New Maps.*

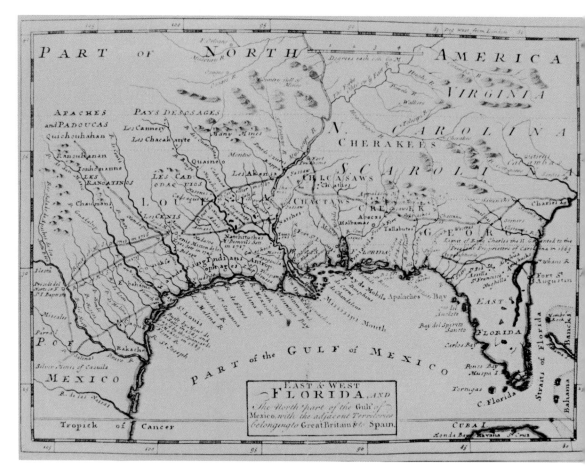

Fig. 2-6. "Florida . . . ," in *Atlas Minor . . .* (1736). *Courtesy Cartographic History Library, University of Texas at Arlington*

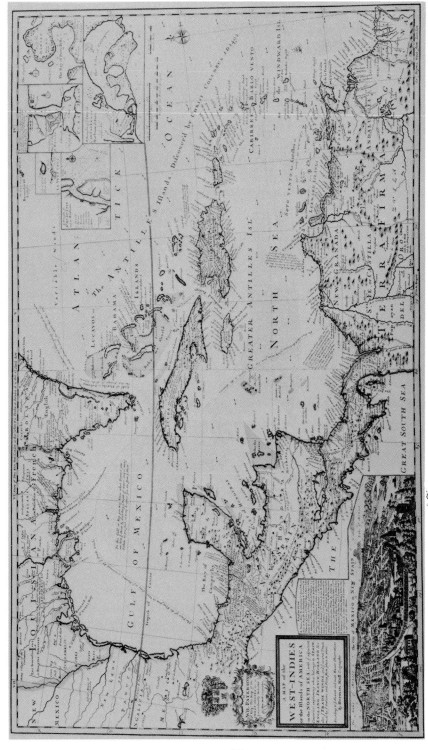

Fig. 2-7. "A Map of the West Indies . . ." in *The World Described . . .* (1715). *From the private collection of Dennis and Judy Reinbartz*

coast across the Gulf of Mexico he traces "The Tract of the Flota from la Vera Cruz to ye Havana, occasioned by ye Trade Winds." Finally, he observes that, "In this Gulf or Bay you may know what Distance you are from ye Shoar by Sounding ye Depth of Water, and as many Fathoms as you find, so many Leagues you are from ye Shoar."[38] Modern charts do not at all bear this out, but such notations also appeared in French on Delisle's maps.

Conclusion

Herman Moll worked at the very dawn of what has been called the "cartographic enlightenment," and it is evident from his mapping of the American Southwest that he was serious about and dedicated to his craft as a geographer. His inaccuracies and inconsistencies were common to the age and the state of the art and were often derived from earlier work by others. He certainly was not atypical, but he was one of the most important European cartographers and imagemakers of the early eighteenth century.

Moll's own definition of cartography appeared many times in his books, but it was perhaps best related in the opening paragraph of the "Advertisement" of the *Atlas Manuale.* . . . It provides an appropriate closure to this examination of Moll and his works:

> The Art of making MAPS and Sea-Charts, is an Invention of such vast use to Mankind, that perhaps there is nothing for which the World is more endebted to studious Labours of Ingenious Men. For by the help of them geography is made plain and easie, the Mariners are directed in fetching us the Commodities of the most distant Parts and by the help of them, we may at home, with Pleasure, survey the several Countries of the World, and be inform'd of the Situation, Distance, Provinces, Cities and remarkable Places of every Nation. To do this with Exactness, was an Art (to be sure) not easily attain'd; it was not one Man, nor one Generation of Men, that could bring it to any reasonable Perfection.[39]

[38] Moll, "A New Map of the North Parts of America claimed by France . . . ," and "A Map of the West Indies . . . ," *The World Described; New Maps.*
[39] Herman Moll, *Atlas Manuale* . . . (London: 1709), p. i.

United States Army Mapping
in Texas, 1848–50

•

ROBERT SIDNEY MARTIN

Jefferson Davis aptly represents both the interest and the ignorance concerning the American Southwest that prevailed in official circles at the close of the Mexican War. As secretary of war from 1853 to 1857, he was to be a staunch proponent of army exploration in the region, and even earlier, as a senator from Mississippi, he was outspoken in his support for such operations. On August 8, 1848, Senator Davis offered a resolution instructing the secretary of war to furnish the Senate a map of Texas and adjacent territories. In speaking for the proposal, Davis explained that "the country referred to in the resolution had been very little explored and . . . we know very little about it."[1]

The little-known territory in question was the Mexican Cession, that vast tract the United States had acquired by the Treaty of Guadalupe-Hidalgo. Davis correctly assessed the state of geographical knowledge concerning the region, for in spite of centuries of Spanish activity, it remained a virtual *terra incognita* to officials in Washington. Perhaps the best example of this ignorance is the official map compiled in 1844 by William H. Emory of the Corps of Topographical Engineers. Designed to furnish information for the congressional debates over the annexation of Texas, the map was a compilation of the best information then available in Washington. But in drafting the map Emory could draw on no official surveys or explorations, for there were none. Instead he assembled information from the various authorities listed on the map's face, including Humboldt, Pike, Austin, and Arrowsmith. Consequently, he frequently had to reconcile conflicting information, and occasionally, unable to determine

[1] *Congressional Globe*, 30th Cong., 1st sess., August 8, 1848, p. 1051.

which data were correct, he was forced to show one feature in two possible locations. The resulting map was a good effort, but it clearly reveals the lack of information concerning the American Southwest at the outset of the Mexican War.[2]

Another example of the dearth of geographical knowledge about the region is the widely circulated map published in 1847 by John Disturnell of New York. The Mexican and American plenipotentiaries used Disturnell's map (actually a plagiarism of a much earlier one) in negotiating the Treaty of Guadalupe-Hidalgo and referred to it in the treaty itself. The map's distortion of the Rio Grande Valley and its displacement of El Paso resulted in a bitter controversy over the boundary and significantly delayed completion of the boundary survey.[3]

United States armed forces operations during the Mexican War added greatly to the store of geographical information, especially concerning California and Mexico, because the military reports usually contained detailed maps of the lines of march. However, most army operations in Texas took place between San Antonio and the lower

[2]William H. Emory, *Map of Texas and the Countries Adjacent: Compiled in the Bureau of the Corps of Topographical Engineers, from the best authorities. For the State Department, under the direction of Colonel J. J. Abert, Chief of the Corps . . .* (Washington: War Department, 1844). For a more detailed discussion of the Emory map, see Robert Sidney Martin and James C. Martin, *Contours of Discovery: Printed Maps Delineating the Texas and Southwestern Chapters in the Cartographic History of North America, 1513–1930* (Austin: Texas State Historical Association, 1981, 1982), pp. 54–55. See also James C. Martin and Robert Sidney Martin, *Maps of Texas and the Southwest, 1513–1900* (Albuquerque: University of New Mexico Press, 1984), pp. 128–29. Thomas W. Streeter, *Bibliography of Texas, 1795–1845*, 5 vols. (Cambridge: Harvard University Press, 1955–60), 2:499–501, Number 1543; Carl I. Wheat, *Mapping the Transmississippi West, 1540–1861*, 5 vols. (San Francisco: Institute for Historical Cartography, 1957–66), 2:190–91, number 478.

[3]John Disturnell, *Mapa de los Estados Unidos de Mejico, Segun lo organizado y definido por las actas varias del Congresso de dicha Republica: y construido por las mejores autoridades* (New York: J. Disturnell, 1847). The best discussion of the Disturnell map remains Lawrence Martin, "Disturnell's Map," in *Treaties and Other International Acts of the United States of America*, ed. David Hunter Miller (Washington: Government Printing Office, 1937), 5:339–70. The essay was reprinted, as edited and abridged by Walter W. Ristow, in *A la Carte: Selected Papers on Maps and Atlases* (Washington: Library of Congress, 1972), pp. 204–21. See also Martin and Martin, *Maps of Texas*, pp. 137–39; Martin and Martin, *Contours of Discovery*, pp. 55–56; and Wheat, *Transmississippi West*, 3:35–37, 45, 51–52, 64–65, 77–78, numbers 507, 540, 556, and 605.

Rio Grande, an area already relatively well-known, and thus contributed little in the way of new geographic knowledge. For example, Capt. George W. Hughes's map, showing the invasion route of Wool's division from San Antonio to Saltillo in 1846, provides much more information about northern Mexico than it does about Texas.[4] Similarly, the otherwise useful map in the "Memoir" of Frederick Adolphus Wislizenus, a German academic who accompanied Doniphan's expedition, offers only a distorted and largely empty depiction of western Texas.[5]

Without question, however, the map published by Jacob De Cordova in 1849 best summarizes the geographical information available about Texas immediately after the Mexican War.[6] Compiled in late 1848 by Robert Creuzbaur, a draftsman in the General Land Office of the State of Texas, and based primarily on the records of that office, the map presents a remarkably detailed and accurate rendering of the area south and east of San Antonio. North and west of that point, however, the data are scarce and the features sparse. The entire Trans-Pecos and the vast claims of Texas on the upper Rio Grande are not even shown on the main map; they appear only on the small inset.[7]

The task of exploring, defining, and defending this unknown territory fell largely to the United States Army, whose Mexican War experiences had underscored the national ignorance of Southwest geography and had reinforced dramatically the need for good communications with the far-flung new territories. The army's Chief of Topo-

[4]George W. Hughes, "Map Showing the Line of March of the Centre Division, Army of Mexico, under the Command of Brigr. Genl. John E. Wool, from San Antonio de Bexar, Texas, to Saltillo, Mexico . . . 1846 . . . ," in *Memoir Descriptive of the March of a Division of the United States Army, under the Command of Brigadier John E. Wool, from San Antonio de Bexar in Texas, to Saltillo, in Mexico*, 31st Cong., 1st sess., 1850, S. Exec. Doc. 32 (serial 558).

[5]Frederick Adolphus Wislizenus, "Map of a Tour from Independence to Santa Fe, Chihuahua, Monterey and Matamoros," in *Memoir of a Tour to Northern Mexico, Connected with Col. Doniphan's Expedition, in 1846 and 1847*, 31st Cong., 1st sess., 1848, S. Misc. Doc. 26 (serial 511).

[6]Robert Creuzbaur, *J. De Cordova's Map of the State of Texas Compiled from the Records of the General Land Office of the State by Robert Creuzbaur, Houston, 1849* (New York: J. De Cordova, 1848).

[7]See Martin and Martin, *Maps of Texas*, pp. 140–41; Martin and Martin, *Contours of Discovery*, p. 57; Wheat, *Transmississippi West*, 3:64, number 603.

graphical Engineers, Col. John James Abert, wrote to an official in the State Department that "unless some easy, cheap, and rapid means of communicating with these distant provinces be accomplished, there is danger, great danger, that they will not constitute parts of our Union."[8] After the Mexican War, the main objective of Abert's office was to locate suitable southwestern routes for a transcontinental railroad joining the Mississippi River with the Pacific provinces. Abert proposed a railroad route from San Diego up the Gila to the Rio Grande, then southeast across Texas by way of El Paso, San Antonio, and Nacogdoches, connecting along the way with the navigational heads of all the Texas rivers. Other individuals favored other routes, and the search for the best route dominated operations of the Corps of Topographical Engineers throughout the American Southwest. Colonel Abert's plans for Texas were not restricted to railroads, however. He also outlined river improvement projects and wagon roads to complete the network of communications.

These activities were complicated by a number of factors, not the least of which was the constant political wrangling over the boundaries of the new territories, particularly Texas. In addition, developing political, commercial, and sectional rivalries ultimately culminating in the Civil War affected developments in Texas. Finally, and perhaps of the greatest immediate importance, thousands of emigrants passed through Texas on their way to the gold fields in California, and the army assumed the new burden of guiding and escorting trains of argonauts across the plains.[9] Thus the army's routine operations in Texas after the Mexican War took place in a hothouse atmosphere, and ac-

[8] Abert to Francis Markoe, May 18, 1849, Letters Sent, Records of the Corps of Engineers, Record Group 77, National Archives, quoted in William H. Goetzmann, *Army Exploration in the American West, 1803–1863* (New Haven: Yale University Press, 1959), p. 209. For a complete discussion of army activity during this period, see Goetzmann, *Army Exploration*, pp. 209–61. Goetzmann's work is the definitive treatment of the Corps of Topographical Engineers and the source for much of this account; except where directly quoted, it will not be cited again.

[9] Concerning the gold rush through Texas, see Ralph P. Bieber, *Southern Trails to California in 1849* (Glendale, California: Arthur H. Clark Co., 1939); Mabelle Eppard Martin, "California Emigrant Roads Through Texas," *Southwestern Historical Quarterly* 28 (1925): 287–301; "From Texas to California in 1849: The Diary of C. C. Cox," *Southwestern Historical Quarterly* 29 (1926): 36–50; Ralph P. Bieber, "Southwestern Trails to California in 1849," *Missouri Valley Historical Review* 12 (1925): 342–75.

tivities that might normally have stretched out over many years were urgently compressed into a few.

On December 10, 1848, Mexican War hero William Jenkins Worth was appointed to command army operations in Texas, which was organized as the Eighth Military Department. Worth's orders were to establish posts on the Rio Grande below San Antonio and along the frontier settlements and to examine the country between San Antonio and the Rio Grande, as far north as Santa Fe.[10] Colonel Abert, who attached great importance to operations in Texas, dispatched one of his most outstanding officers, Brevet Lt. Col. Joseph E. Johnston, for service under Worth. A war hero in his own right and later an important Confederate general, Johnston commanded four young lieutenants, whom he kept busy exploring routes for wagon roads and surveying rivers.

The Corps of Topographical Engineers was not the only army group exploring the Eighth Military Department. The regular engineers were responsible for constructing and maintaining military outposts. The Quartermaster Corps wanted to develop adequate wagon roads for moving supplies to the widely dispersed forces of the region. Line officers also played an important role, most notably Capt. Randolph B. Marcy of the infantry. Assisting the army's explorations were Texas Rangers John Coffee Hays and John S. "Rip" Ford, veterans of the Mexican campaign, and a number of civilians, including Richard A. Howard and J. F. Minter.

In the brief period between 1848 and 1850, these individuals and others undertook numerous surveys in Texas. Their official reports often included maps, which were gathered in central offices, compared and collated, and ultimately compiled into larger and increasingly more accurate and detailed depictions of the territory. Although the expeditions themselves have been the subject of an impressive body of scholarship, the cartographic records which they generated have, for the most part, escaped the scrutiny of historians.[11]

[10] A. B. Bender, "Opening Routes Across West Texas, 1848–1850," *Southwestern Historical Quarterly* 37 (1933): 119.

[11] In addition to Goetzmann's *Army Exploration*, there are several other important studies of these expeditions. Perhaps the earliest discussion is Gouverneur K. Warren, "Memoir to Accompany the Map of the Territory of the United States from the Mississippi to the Pacific Ocean . . . ," in *Reports of Explorations and Surveys, to Ascertain the Most Practicable and Economical Route for a Railroad from the*

The earliest postwar expeditions across the unknown regions of Texas were not strictly military endeavors, but they bore directly on those which followed. The first such expedition, organized by a group of San Antonio citizens and led by former Texas Ranger John Coffee Hays, was an attempt to establish a wagon route between San Antonio and Chihuahua and thus divert the lucrative Santa Fe trade. For three months in the fall of 1848, Hays and his party of seventy-five men floundered through the Big Bend region on the Rio Grande. Exhausted, they finally reached Presidio del Norte, across from present Presidio, Texas, where, still short of their goal, they turned back for San Antonio. In spite of the abject failure of their mission, Hays reported optimistically of discovering a route across the Trans-Pecos to El Paso and beyond. His report contained no map, but the editor of the *Telegraph and Texas Register* published the report with other documents and a crude sketch of the territory, showing a vague route spanning a distorted West Texas and connecting with Cooke's Wagon Road at the Rio Grande.[12]

In February, 1849, shortly after Hays's return, General Worth commissioned the federal Indian Agent for Texas, Robert Simpson Neighbors, "to examine the country between the Pecos and El Paso del Norte, and ascertain whether a practicable route could be found for passing troops."[13] Neighbors recruited another former Ranger, John Salmon "Rip" Ford, to accompany him. Together they discovered a route that came to be known as the Upper Road. It passed from the head of the Concho River near present San Angelo across to the Pecos at Horsehead Crossing, up that river to Deleware Creek, and then to El Paso by way of Guadalupe Pass and the Hueco Mountains.[14]

Mississippi River to the Pacific Ocean, 1853–6," 33d Cong., 2d sess., 1855, *S. Exec. Doc.* 78, XI (serial 768) or *H. Exec. Doc.* 91, XI (serial 801), pp. 56–57, 60–62. See also William Turrentine Jackson, *Wagon Roads West: A Study of the Federal Road Surveys and Construction in the Transmississippi West* (Berkeley: University of California Press, 1952), pp. 36–47, and Bender, "Opening Routes," pp. 116–35.

[12]Jackson, *Wagon Roads*, p. 37; Martin, "California Emigrant Roads," p. 291; Bender, "Opening Routes," pp. 117–19; Bieber, "Southwestern Trails to California," p. 353; James Kimmins Greer, *Colonel Jack Hays: Texas Frontier Leader and California Builder* (New York: Dutton, 1952), pp. 217–26; *Telegraph and Texas Register* (Houston), January 25, 1849.

[13]"The Report of the Expedition of Major Robert S. Neighbors to El Paso in 1849," *Southwestern Historical Quarterly* 60 (1957): 528.

[14]Kenneth F. Neighbours, "The Expedition of Major Robert S. Neighbors to

Neighbors's report apparently contained a map, but that has not been located.[15] The expedition nevertheless bore immediate cartographic fruit, for Ford's report was incorporated into a pamphlet published in 1849 by the enterprising draftsman Robert Creuzbaur, who had earlier compiled De Cordova's map. Entitled *Route from the Gulf of Mexico and the Lower Mississippi Valley to California and the Pacific Ocean, Illustrated by a General Map and Sectional Maps, with Directions to Travelers,*[16] this forty-page work was clearly designed to capitalize on the gold rush. Creuzbaur advocated a Texas route based in part on the records of the General Land Office, where he was employed, and in part on Ford's report, which Creuzbaur reproduced in full.

As the title indicates, Creuzbaur's work includes a general map, supplemented by four sectional ones. The title of the general map is descriptive: *A Map to Illustrate the Most Advantageous Communication from the Gulf of Mexico and the Mississippi Valley to California and the Pacific Ocean.*[17] It exhibits the outward track of Neighbors and Ford, skirting to the north of the Davis Mountains (here called "Sierra Pah-Cut"), striking the Rio Grande above Presidio, and proceeding on to El Paso. It also shows their return route via Hueco Tanks, Guadalupe Pass, and the Pecos—the route Neighbors and Ford recommended. Creuzbaur's "Proposed Route" cuts north from the head of the Concho to a higher crossing of the Pecos, proceeds through Guadalupe Pass, and then strikes out overland to reach the Rio Grande at San Diego, some miles above Doña Ana. The map is very accurate for the period, and it is hardly surprising that the eastern part of Texas bears a remarkable resemblance to De Cordova's map.

The first of the pamphlet's sectional maps, showing Texas west of Austin, was "compiled mostly from the journal & notes taken by Dr. John S. Ford of his exploring expedition in company with Mr.

El Paso in 1849,"*Southwestern Historical Quarterly* 58 (1954): 36–59; Kenneth F. Neighbours, *Robert Simpson Neighbors and the Texas Frontier, 1836–1859* (Waco: Texian Press, 1975), pp. 68–86; John Salmon Ford, *Rip Ford's Texas* (Austin: University of Texas Press, 1963), pp. 115–29; W. J. Hughes, *Rebellious Ranger: Rip Ford and the Old Southwest* (Norman: University of Oklahoma Press, 1968), pp. 57–73; "Opening Routes to El Paso, 1849," *Southwestern Historical Quarterly* 48 (1944): 262–72.

[15]Neighbors, "Report of the Expedition," p. 529.

[16](New York: H. Long and Brother; Austin: Robert Creuzbaur, 1849).

[17]See Wheat, *Transmississippi West,* 4:75–77, number 597; Martin, "California Emigrant Roads," p. 297.

Robert S. Neighbors." This detailed sketch shows the Neighbors-Ford route in its entirety, along with "suggested improvements." Also shown in the Trans-Pecos is "Capt. May's Trail," the track of the first group of argonauts to cross the region earlier that year, and along the Rio Grande in the Big Bend is "Lieut. Whiting's Trail."[18]

On February 9, 1849, Lt. William Henry Chase Whiting of the Corps of Engineers was ordered by General Worth to explore the route that John Coffee Hays had recommended and "to ascertain if there be a practicable and convenient route for military and commercial purposes between El Paso and the Gulf of Mexico . . . ." Whiting was assisted by Topographical Engineer William Farrar Smith and guided by Texan scout Richard A. Howard, whom both officers later praised for his judgment and knowledge of the terrain. Although their explorations virtually coincided with those of Neighbors and Ford, the engineers discovered a more southern route than the civilians'. It became known as the Lower Road.[19]

The reports and journals of Whiting and Smith make interesting reading. They left San Antonio rather hurriedly on February 12, 1849, proceeding to the head of the San Saba River, then crossed over to the Pecos, going three days without water. Reaching the Pecos near its confluence with Live Oak Creek, they proceeded north along the river about forty miles before striking out west via Comanche Springs, near present Fort Stockton. They struck Limpia Creek and followed it to its source in the Davis Mountains, which they called the Diablo, and en route discovered and named Wild Rose Pass. Narrowly escaping from a large and hostile band of Apaches, the party proceeded south to Leaton's Fort, near Presidio del Norte. Between there and El Paso they encountered some difficulty traversing the canyons of the Rio Grande. After a brief rest they returned via a more direct route through the Davis Mountains to the Pecos; in order to avoid the dry camps they had endured on the outward march, they continued south along the Pecos for some miles before crossing over to

[18]"Map of a route from Austin-City to Paso del Norte &c. Compiled, mostly, from the journal & notes taken by Dr. John S. Ford of his exploring expedition in company with Maj. Robert S. Neighbors in March, April & May, 1849. Robert Creuzbaur fecit."

[19] *Report of Lieut. W. H. C. Whiting, Corps of Engineers, of the exploration of a new route from San Antonio de Bexar to El Paso*, 31st Cong., 1st sess., 1850, *H. Exec. Doc. 5* (serial 569), p. 292.

the Devil's River, which they in turn followed almost to its mouth. They then turned almost due east across the Leona, Nueces, and Frio rivers, reaching San Antonio on May 24, 1849.[20]

Both young lieutenants reported that their return route was satisfactory for wagons. Whiting described in detail the many difficulties under which they had labored, noting that "in so extensive a country, and so little known, it cannot be said that neither other nor better routes do not exist. Much remains to be known. The geography and geology of the whole region is yet to be settled." He pointed out that the expedition was not a scientific one, "for we had no time to supply either books, instruments, or maps." He included with his report "a rough sketch of the march, made by the compass, and of course only an approximation of the geography of the country." This sketch has not been located. Curiously enough, the topographical engineer, Smith, apparently made no map to accompany his report, but he did append a list of the latitudes of key points, including El Paso. Both men found that their timepieces were too unreliable to permit any calculations of longitudes.[21]

The explorations by Whiting, Smith, Neighbors, and Ford resulted in not one but two recommended routes from San Antonio to El Paso, the Upper and Lower roads. To determine which route was better, the chief topographical engineer in Texas, Colonel Johnston, detailed Lt. Francis T. Bryan to "make a reconnaissance of the route . . . to El Paso del Norte—the same lately passed over by Major Neighbors, Indian Agent." Bryan was instructed "to obtain, with perfect accuracy, the best information in reference to a permanent military road from the Gulf of Mexico to El Paso." He was ordered "to be particular in your examinations and observations" and to make a de-

---

[20]Ibid., pp. 281–93; Smith to Johnston, May 25, 1849, Reports of the Secretary of War, with Reconnaissances, 31st Cong., 1st sess., 1850, *S. Exec. Doc. 64* (serial 562), pp. 4–7 (*S. Exec. Doc. 64* hereafter cited as *Reports*); "Whiting's Diary," *Publications of the Southern History Association* 6 (1902): 283–94, 389–99, 9 (1905): 361–73, 10 (1906): 1–18, 78–95, 127–40; "The Journal of Lieutenant W. H. C. Whiting," in *Exploring Southwestern Trails, 1846–1854*, ed. Ralph P. Bieber and A. B. Bender (Glendale, California: Arthur H. Clark Co., 1938), pp. 241–350. See also Warren, "Memoir," p. 60; Jackson, *Wagon Roads*, pp. 39–40; Bender, "Opening Routes," pp. 121–23; and Bieber, "Southwestern Trails to California," pp. 353–54.

[21]"Report of Lieut. Whiting," pp. 292–93; *Reports*, p. 7; "Whiting's Diary," 6:289, 10:137.

tailed report "in order that a comparison may be drawn between this and the route recently explored by Lieutenants Whiting and Smith."[22] Leaving San Antonio on June 14 with thirty men, Bryan took only forty-six days to reach El Paso. He measured the distances along the way with an odometer and included a table detailing them in his report. He concluded that the route traveled "presents no obstruction to the easy passage of wagons," and his only recommendation was to sink several wells, which could shorten the route by several miles.[23]

Meanwhile, another force under the command of Colonel Johnston himself was reexamining the Lower Road. They left San Antonio on June 13, accompanying Maj. Jefferson Van Horn's battalion of the Third Infantry and a long train of supply wagons en route to El Paso. Lieutenant Smith and scout Howard, just returned from El Paso, guided the expedition. In two places the party deviated from Smith and Whiting's trail: near the first crossing of the Devil's River, and near the Davis Mountains, where they failed to find a way to the north. Because it was encumbered with wagons and infantry, the expedition took almost three months to make the trek, arriving in El Paso on September 8, 1849. Meanwhile, as ordered, Lieutenant Bryan waited there for his commander to arrive.

While resting their mounts in preparation for the return march, Johnston and his subordinates surveyed the Rio Grande Valley from Ysleta to Doña Ana and unsuccessfully attempted to find a pass through the Sacramento Mountains. The party set out for San Antonio on October 11, 1849, following the Upper Road. When they reached the Pecos, Johnston decided to try a more southerly route back to San Antonio. They therefore proceeded down the Pecos and returned the rest of the way by the Lower Road. Along the way Johnston and Howard diverged from the rest of the party in an unsuccessful attempt to find a more direct line from the head of the Devil's River to the Nueces.[24]

Johnston's explorations in the summer and fall of 1849 were extremely important. As William H. Goetzmann notes, "by means of this extended reconnaissance the chief Topographical Engineer in Texas

---

[22] *Reports*, pp. 25–26.

[23] Ibid., pp. 14–24, quote from p. 23.

[24] Ibid., pp. 26–28; Warren, "Memoir," p. 61; Jackson, *Wagon Roads*, pp. 40–42; Bender, "Opening Routes," pp. 123–25.

had been able personally to survey what was to be the most important supply route for the outer chain of frontier defense posts, and he had also been able to gain some idea of the suitability of the terrain for a railroad."[25]

Throughout these reconnaissances the topographical engineers had measured the trail distances with odometers and had taken numerous astronomical observations with a sextant. After returning to San Antonio, Johnston and his officers gathered their data, drafted their reports, and prepared a general map covering all of their activities. This "Sketch of Routes from San Antonio de Bexar to El Paso del Norte, 1849" is the first map based on actual surveys of the region. It is a document of enormous importance, displaying both the Upper and Lower roads as well as all the permanent sources of water so important in this arid land. A note by Colonel Johnston states that "The lower road from San Antonio to the Frio is taken from the field notes in the Bexar County Land Office. The upper road from San Antonio to the San Saba River is copied from Decordova's Map of Texas."[26]

The map also details Johnston's explorations in the valley of the Rio Grande around El Paso, and in the Sacramento Mountains. Also shown are the abortive tracks north of the Davis Mountains (here called the Apache Mountains) and the divergent route taken by Johnston and Howard between the Devil's River (here still called the San Pedro) and the Nueces. Amid the maze of Indian trails between the Davis Mountains and the Rio Grande and between the Pecos and the Devil's River, the return route of Lieutenants Smith and Whiting is shown where it differs from Johnston's. Not shown is Whiting and Smith's outward track, by way of Presidio, nor that of Ford and Neighbors, skirting the Davis Mountains—probably because they were not recommended routes. Also absent from the sketch is any trace of the Rio Grande between Presidio and the mouth of the Devil's River, which had not yet been explored.

Johnston's report relates the map to all of the army's operations

[25]Goetzmann, *Army Exploration*, p. 233.

[26]Warren, "Memoir," p. 61; "Sketch of Routes from San Antonio de Bexar to El Paso de Norte. 1849. Scale of ten miles to one inch. By Lt. Col. J. E. Johnstone Top. Engrs. Lt. W. F. Smith Top. Engrs. Lt. F. T. Bryan Top. Engrs. Mr. R. A. Howard," Record Group 77, National Archives, US 143-1.

in the region. "By referring to the accompanying sketch," he wrote, "you will see that both the routes now used very far exceed the direct distance to El Paso." He strongly urged that, in the interest of shortening these routes, further extensive explorations be conducted in the Trans-Pecos. The advantages of such explorations included opening the river to navigation, improving communications all along the frontier, and promoting settlement in the valley of the Rio Grande.[27] Johnston's recommendations were to bear fruit, but that lies outside the scope of this study.

While these topographical engineers had been probing the wastelands of the Trans-Pecos, other officers were simultaneously pursuing similar activities both to the north and the south of them. In January, 1849, Lt. Nathaniel Michler, assisted by Lieutenant Bryan, reconnoitered Aransas and Corpus Christi bays, seeking the optimal site for a supply depot on the coast. These two officers also laid out a road from Corpus Christi to San Antonio via San Patricio, and the following month they made a reconnaissance of the lower road from San Antonio to Presidio de Rio Grande, via Fort Inge on the Leona River. In May, Michler alone examined the road from San Antonio to Port Lavaca, and in June and July he reconnoitered the route from Corpus Christi to Fort Inge, by way of the Nueces, Frio, and Leona rivers.[28] This last venture was apparently the only one for which a formal report and map were prepared.[29] This sketch shows Michler's route in detail and notes the location of pertinent features and landmarks, as well as the courses of streams.

Meanwhile, far to the north, other officers had been equally hard at work. On April 2, 1849, Capt. Randolph B. Marcy was ordered to escort a large group of California-bound emigrants as far as Santa Fe, following the south side of the Canadian River all the way. Marcy's main object was to defend the emigrants from Indian attacks, but he was also to ascertain, insofar as possible, the best route to New Mexico and California. Joining the party a short distance along

[27] *Reports*, p. 27.

[28] Warren, "Memoir," pp. 61–62; Bender, "Opening Routes," p. 130.

[29] *Reports*, pp. 7–13; "Sketch of a Reconnoissance [*sic*] from Corpus Christi to the Post on the Leona, By Lieut. N. Michler," Record Group 77, National Archives, US 143-3.

the trail was one of Abert's most able young topographical engineers, James Hervey Simpson, who had orders to survey the route all the way to California.[30]

The expedition reached Santa Fe on June 28, 1849, and within the week Simpson received new orders, directing him to stop there and immediately draft a report and map of the route that far. This report, dated August 13, 1849, includes four detailed maps showing the whole route traversed. Although these maps show little but the course of the river and the expedition's track along it, an area already relatively well known from prior explorations, Simpson had used a compass and surveyor's chain and had made frequent astronomical observations with a sextant and chronometer. Thus the maps are extremely accurate in both the locations of principal points and the distances between them, marking an important advance in the cartography of the region.[31]

While Simpson was compiling data in Sante Fe, Marcy, as instructed, made additional reconnaissances. He advanced south along the Rio Grande to Doña Ana and there turned east "with the intention of blazing a trail across Texas that would form a more direct connection with Cooke's wagon route . . . west."[32] Marcy left the Rio Grande, crossed the Organ Mountains, and struck Lieutenant Bryan's trail just before entering the Hueco Mountains. He followed Bryan's route as far as Emigrant Crossing on the Pecos, then turned northeast toward Preston on the Red River. He traversed the southern rim of the Llano Estacado via the Monahans sand dunes, the Big Spring at the head of the Colorado, and the heads of the Brazos and Trinity rivers. Marcy reported that this return route was even better for wagons than his outward one had been. It also, unbeknownst to Marcy,

---

[30] Jackson, *Wagon Roads*, pp. 24–29; Warren, "Memoir," p. 56; Bieber, "Southwestern Trails to California," pp. 357–63.

[31] Jackson, *Wagon Roads*, pp. 24–29; Warren, "Memoir," p. 56; Bieber, "Southwestern Trails to California," pp. 357–63; James H. Simpson, "Report of Exploration and Survey of Route from Fort Smith, Arkansas, to Santa Fe, New Mexico," *S. Exec. Doc.* 12 (serial 554), 31st Cong., 1st sess., 1850, or *H. Exec. Doc.* 45 (serial 577); James H. Simpson, *Map of Route pursued by U. S. Troops from Fort Smith, Arkansas, to Santa Fe, New Mexico, via south side of Canadian River in the year 1849* . . . (Washington, D.C., 1850). See also Wheat, *Transmississippi West*, 3:16, 286, number 640.

[32] Goetzmann, *Army Exploration*, p. 217.

coincided with Colonel Abert's plan for a road close to the heads of the Texas rivers.[33]

The map which accompanies Marcy's report portrays the entire vast loop of his expedition from Fort Smith, Arkansas, to Santa Fe and back. The outward track he shows is, of course, very similar to Simpson's, but this map is nowhere as detailed or precise. Marcy, an infantry officer, was supplied with neither instruments for astronomical observations nor chains for measuring precise distances. His mileages along the way were provided by an odometer.[34] But as the first comprehensive depiction ever published for that portion of Texas, his map proved valuable to subsequent parties of emigrants following the route he had explored. As Goetzmann notes, the maps of both Simpson and Marcy "filled a gap in the knowledge of western geography."[35]

Even while Marcy was on the trail somewhere in Texas, Lieutenant Michler in San Antonio received new orders to examine the same country as the infantryman and to locate "the route from the upper valley of the south branch of the Red River to the Rio Pecos."[36] Michler decided it would be best to travel from north to south, so he proceeded from San Antonio to Fort Washita in present Oklahoma, passing through Austin, Navarro, Dallas, and Preston along the way. Marcy arrived at Fort Washita just as Michler was preparing to leave, and the two officers compared notes. Michler's route lay to the north of Marcy's until he passed the Brazos, but from near there to the Pecos the trails were identical. Michler returned to San Antonio by the Upper Road, arriving January 28, 1850, and documented his reconnaissance with a report and a map. Like Simpson, Michler had laid out his road with compass and chain, but his report makes no mention of astronomical observations.[37]

---

[33] Warren, "Memoir," p. 57; "Report of Capt. R. B. Marcy," *H. Exec. Doc. 45* (serial 577), 31st Cong., 1st sess., 1850; also *Reports*, pp. 169–233.

[34] Randolph B. Marcy, "Topocraphical [sic] Map of the Road from Fort Smith, Arks. to Santa Fe, N.M and from Dona [sic] Ana, N.M, to Fort Smith . . ." *H. Exec. Doc. 45* (serial 577), 31st Cong., 1st sess., 1850; Wheat, *Transmississippi West*, 3:12–14, number 681; Warren, "Memoir," p. 57.

[35] Goetzmann, *Army Exploration*, p. 218.

[36] *Reports*, p. 29; "Routes from the Western Boundary of Arkansas to Santa Fe and the Valley of the Rio Grande," *H. Exec. Doc. 67* (serial 577), 31st Cong., 1st sess., 1850, p. 2.

[37] Michler's report in "Routes from the Western Boundary of Arkansas"; "Sketch

In forwarding Michler's map to Abert, Colonel Johnston pointed out that the survey demonstrated conclusively "that there is no obstacle to the construction of a road of any sort from the neighborhood of Fulton, Arkansas, to the Pecos, in the direction of El Paso." The rest of the route had already been determined. Abert was no doubt pleased to find a route so favorable to his plans, and in forwarding the report to the secretary of war he expanded Michler's map to incorporate "other routes in the same direction, previously examined and duly reported by Colonel Johnston."[38]

A map fragment answering Abert's description is found in the records of the War Department in the National Archives.[39] It looks strikingly like Johnston's earlier compilation, with the added explorations by Michler along the Nueces and between the Pecos and the Red; no mention is made of Marcy's expedition. Along the San Saba a trail which cuts off from the Upper Road is labeled "Emigrant's Route." The map remains something of an enigma, for unlike the earlier compilations, it is printed, not manuscript, but there is no record of congressional authorization to publish such a map at this time. It is unclear whether this is the map Abert mentioned in his letter, but it does represent the cartography of the region as it had developed to that date.

While Michler was engaged in his survey, an inspection of another sort was being conducted in another part of the state. In October, 1849, the Commander of the Eighth Military Department ordered Lieutenant Whiting to examine the entire military frontier of West Texas from the Rio Grande to the Red River. Whiting was to recommend the disposition of existing posts, the construction of new ones, and the development of roads to connect them. He was also instructed to observe the general nature of the country, the availability of stone, timber, and water, and the status of the civilian settlements he found.

---

of a Reconnaissance from Fort Washita to the Pecos, By Lt. N. Michler. Topog; Engre:.," Record Group 77, National Archives, U.S. 143-4; Warren, "Memoir," p. 61; Jackson, *Wagon Roads*, p. 43; Bender, "Opening Routes," pp. 131–34.

[38] Johnston to Abert, March 10, 1850, and Abert to G. W. Crawford, April 22, 1850, in *Routes from Arkansas to the Rio Grande*.

[39] "Reconnaissances of routes from San Antonio de Bexar to El Paso del Norte, etc. by Bvt. Lt. Col. J. E. Johnston, T. Engrs. Lt. W. F. Smith, Lt. F. T. Bryan, Lt. N. H. Michler, 1849. Engrd. by J. Mediary," Record Group 77, National Archives, U.S. 143-5.

His comprehensive report is an important document, describing the entire Texas frontier at an early stage of its evolution, and his inspection map accurately depicts much of this territory for the first time.[40]

The army was interested in far more than just wagon roads and frontier forts. Colonel Abert's plan for regional development included river resources as well. In this he was firmly allied with the commercial and political interests in the state, who steadily petitioned Congress for river surveys by the army. At the earliest possible date, therefore, the topographical engineers set out to examine the important rivers in detail. The first to claim their attention was the Colorado.

In early 1850, Lt. W. F. Smith and civilian surveyors Richard A. Howard and J. F. Minter were detailed to survey the Colorado from Austin to its mouth. Their report, submitted in April, 1850, included a detailed map showing the rafts, snags, sandbars, and falls which impeded navigation along the river's course. In forwarding the report to Abert, Johnston estimated that it would cost $56,000 to remove these obstacles and open the river to navigation. He noted that this cost should be compared to an annual savings of $20,000 in transportation costs were the river to be opened.[41]

By the middle of 1850 this flurry of army activity had produced a surprising array of geographical information about Texas. For the most part, however, this information remained hidden in manuscript maps and reports of limited circulation. On June 8, 1850, the Senate—at the instigation of Thomas Jefferson Rusk, Jefferson Davis, and a number of others—instructed the secretary of war to furnish copies of all the reports and maps not already communicated to the Senate.[42] Colonel Abert forwarded seven reports, embracing all of the army's topographical activities in Texas to that date. With these reports he included a single map, incorporating all the data from these surveys and depicting all of the expedition routes between San An-

[40]Bender, "Opening Routes," pp. 126–27; "Report of Lieutenant W. H. C. Whiting's reconnaissance of the western frontier of Texas," *Reports*, pp. 235–50; "Sketch of a Reconnoissance [*sic*] of the Military Frontier of Texas from the mouth of the False Washita to Eagle Pass. By Lieut. W. H. C. Whiting, U.S. Engrs.," Record Group 77, National Archives, Q34.

[41]*Reports*, pp. 39–40; "Map of the Colorado River from Austin to its mouth. Drawn by Lieut. Wm. F. Smith Topl. Engrs. R. A. Howard Esq.," Record Group 77, National Archives, Q51.

[42]*Congressional Globe*, 31st Cong., 1st sess., June 8, 1850, p. 1170.

tonio and El Paso, along the Nueces and the Leona, from Fort Washita to the Pecos, and along the Colorado River.[43]

The reports and map reached the Senate on July 25, 1850, and immediately caused some controversy. Jefferson Davis pointed out that the general map which he had requested in 1848, showing all army operations in Texas and adjacent territories, was still being prepared, and he hoped that it might be presented to the Senate soon. Davis implied that this comprehensive map might obviate the more particular one presented with the reports, and he therefore objected to printing it until the two maps could be compared. The matter was referred to the Committee on Printing, which reported back the following day in favor of publishing the documents. Then there was some discussion whether the documents supplied were actually those requested, and Davis again took the opportunity to promote the larger map on which he had rested his hopes.[44] The matter was tabled until August 10, 1850, when Davis himself submitted a resolution to print the documents Abert had submitted, along with additional reports by Samuel G. French of the Quartermaster Corps, who had accompanied a number of expeditions, and Lieutenant Simpson, who had completed his journey to the Navajo country. These were to be printed "with the map and such of the accompanying sketches as in the opinion of the chief of Topographical Engineers may be necessary." The resulting grab bag of documents was further expanded to include Whiting's and Marcy's reports and ultimately included two maps—one showing Simpson's foray into the Navajo country and the other summarizing all the explorations in Texas.[45]

This latter map, based on the one Abert submitted in July, also showed Whiting's route of inspection and Marcy's return route from the Rio Grande to Fort Smith, Arkansas.[46] The map therefore stands

---

[43] Ibid., July 24, 1850, pp. 1446–47; *Reports*, pp. 2–3; "Reconnaissance of routes from San Antonio de Bexar to El Paso del Norte. &c. &c. by Bvt. Lt. Col. J. E. Johnston, T. Engrs. Lt. W. F. Smith, Lt. F. T. Bryan, Lt. N. H. Michler, 1849," Record Group 77, National Archives, U.S. 143-2.

[44] *Congressional Globe*, 31st Cong., 1st sess., July 24, 1850, pp. 1446–47, 1456.

[45] Ibid., August 10, 1850, p. 1558; see *Reports*.

[46] Joseph E. Johnston, "Reconnoissances [sic] of Routes from San Antonio de Bexar to El Paso del Norte, &c. &c., by Bvt. Lt. Col. J. E. Johnston, T. Engrs. Lt. W. F. Smith, Lt. F. T. Bryan, Lt. N. H. Michler, 1849. Including the Reconnaissance of Lt. W. H. C. Whiting, U.S. Engrs. 1849.," in *Reports*.

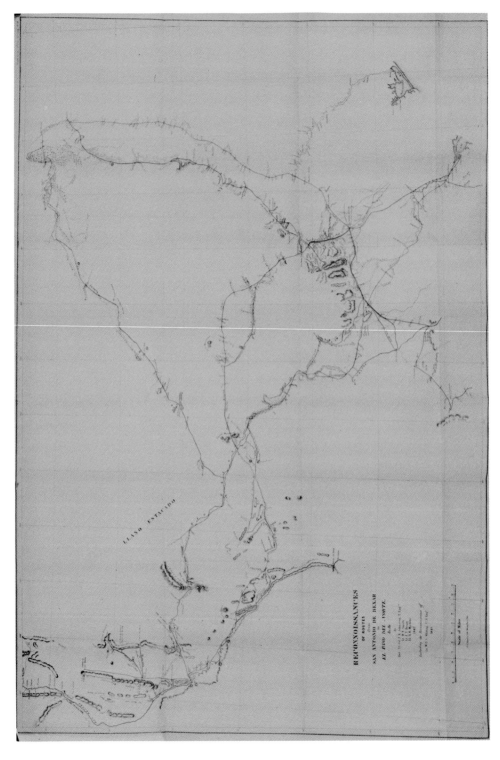

Fig. 3-1. "Reconnoissances [*sic*] of Routes from San Antonio De Bexar to El Paso Del Norte," by Joseph E. Johnston et al. (original size 24 in. by 36 in.). *Courtesy Cartographic History Library, University of Texas at Arlington*

as a single document summarizing all army exploration in Texas between 1848 and 1850 and all of the important discoveries made during this brief period.

The great general map that Jefferson Davis had requested some two years earlier was finally delivered to the Senate on September 6, 1850.[47] In the accompanying letter, Colonel Abert admitted that he had interpreted the Senate resolution broadly. As a result, the map exhibited "not merely the lines of military operations within the extent of the country indicated, but the whole country . . . between the Mississippi and the Pacific." It presented "in one connected view the whole of that extensive region, and in accordance with the best authorities which could be obtained." The colonel pointed out that compiling the map "has been a work of great labor, from the difficulty in procuring authorities, and in the difficulties in reconciling conflicting authorities." He also suggested that the map be considered only a preliminary attempt: "I have no doubt it will hereafter be improved, or that it would be improved if longer retained in this office, from information occasionally received. But believing it to be the best which can be compiled from the information now in the possession of this office, I have considered it proper to bring the work to a close, and to send the map, for the use of the Senate, as it now is."[48]

Davis was more than satisfied by the large and impressive map. He pronounced it to be "the first very valuable contribution which our country will have made to physical geography. . . . [It is] something which will be very valuable at home and greatly sought after abroad."[49] The Senate duly ordered the map to be reduced from the enormous eight-foot-square original and then printed. Although it bears the date of 1850, the process of reducing it probably took months,

[47] *Congressional Globe*, 31st Cong., 1st sess., September 6, 1850, p. 1767; "A Map of the United States and their Territories between the Mississippi and the Pacific, and a part of Mexico. Compiled in the Bureau of the Corps of Topographical Engineers, under a resolution of the U. S. Senate from the best authorities which could be obtained; by Brevet Lt. Col. J. McClellan, of the Corps of Topographical Engineers with corrections from more recent authorities by Bvt. Capt. L. Sitgreaves T.E. as assigned and approved, September, 1850, J. J. Abert, Col. Corps. T. Engrs.," Record Group 77, National Archives, U.S. 150.

[48] *Report from the Colonel of Topographical Engineers* . . . , 31st Cong., 1st sess., 1850, *S. Exec. Doc. 73* (serial 562), p. 2.

[49] *Congressional Globe*, 31st Cong., 1st sess., September 8, 1850, p. 1767.

and the map most likely was not sent to the printer before the first of the year. Nevertheless, the published map depicts army activities through the summer of 1850.[50] As Carl Wheat notes, the map is a "synthesis of the many and varied activities of the U. S. Army carried on in the West since the onset of the Mexican War."[51]

Ironically, this map, which was primarily to show the topographical operations of the army in Texas, portrays these explorations in considerably less detail than the reconnaissance map published earlier that summer. The operations map shows actual routes of the expeditions in Texas but very little about the terrain they traversed. As far as the cartography of Texas is concerned, therefore, the more detailed map by Johnston and his officers remains the substantially more important document.

The army did not, of course, suddenly halt its operations in Texas in midsummer 1850. There remained much yet to discover, and there were more explorations before the year drew to a close. For the most part, however, these subsequent efforts were for the United States–Mexican Boundary Survey and therefore belong to another chapter of the cartographic history of Texas. Even later, there were more detailed explorations under the Pacific Railroad Survey, but the Civil War halted all such operations. The U.S. Army's explorations in Texas during the brief period from 1848 to 1850 remain the important first step in this lengthy effort to lift the veil of obscurity from a barren and challenging land.

[50] The reduced manuscript map is "Map of the United States and their territories between the Mississippi and the Pacific Ocean and of part of Mexico compiled in the Bureau of the Corps of Topogl. Engrs., under a resolution of the U. S. Senate; from the best authorities which could be obtained. 1850," Record Group 77, National Archives, U.S. 152-2. The published version is *Map of the United States and their Territories between the Mississippi and the Pacific Ocean; and of part of Mexico. Compiled in the Bureau of Topogl. Engrs. under a Resolution of the U. S. Senate. From the best authorities which could be obtained. 1850. Engraved by Sherman and Smith. New York* (Washington: War Department, 1850).

[51] Wheat, *Transmississippi West,* 3:110, number 696.

Images of the Southwest in Nineteenth-Century American Atlases

•

JUDITH A. TYNER

The nineteenth century was a period of great change for the United States. Newly emerged from colonial status, the young country was striving to develop new industries in order to be economically as well as politically independent. Coincident with commercial expansion was territorial expansion. Population increased rapidly in the nineteenth century, and the westward migration began. Maps of the newly explored and settled areas were needed.

Thus the beginning of the nineteenth century marks the beginning of American commercial cartography and American atlas production. Prior to the 1790s, atlases used in this country were European, most commonly British. Even the famous Arrowsmith and Lewis atlas that was published in this country was essentially English in origin, although the United States maps were made here and were largely the work of Samuel Lewis.

All of that began to change with the Louisiana Purchase. The newly acquired territory was virtually *terra incognita* — a vast, nearly blank space on maps which were not accurate even in the longitudinal dimension. The expeditions of Long, Pike, Lewis and Clark, and the Army Topographical Engineers provided both government and private cartographers with an abundance of mappable data.

These expeditions also helped to shape the East Coast's images of the West. In *Exploration and Empire,* William Goetzmann has noted that the nineteenth-century confrontation with the unknown was a uniquely western phenomenon, but it created a series of images in the eastern centers of dominant culture and conditioned popular attitudes and public policy concerning the new lands. Impressions which emerged from charts, reports, adventure novels, photographs, school geographies, and even children's books became crucial in shaping the long-range destiny of the West.[1]

Furthermore, images of the West and Southwest have changed through time, from the earliest Spanish explorers even to the present. The area across the Mississippi has been perceived as everything from El Dorado to the Great American Desert, from fruited plain to barren wasteland. These images have changed spasmodically rather than progressively; as explorers' discoveries filled in the gaps of knowledge like a jigsaw puzzle, the images changed at different paces, sometimes leaping forward and sometimes regressing.

In order to understand the influence of nineteenth-century American atlases on the public's images of the Southwest, it is first necessary to review the different ways that images are formed. When one looks at a scene, an instant image forms in the mind, and each viewer will have a different image of the same scene. When an explorer recalls a scene and writes about it in his journal, he is using a recalled image to prepare a message image. Message images, whether in written or map form, undergo distortion before transmission, during transmission, and yet again when the receiver converts the image to a new form of the instant image.[2]

It is my premise that, in part, the educated American's early images of the nineteenth-century Southwest were reflected in and transmitted by the atlas maps. Most people did not have wall maps or the sheet maps of various expeditions, but they probably had access to an atlas of some kind. Although the monumental or "coffee table" atlases were quite expensive—thirty dollars for Tanner's *New American Atlas* of 1822—there were numerous inexpensive school atlases which accompanied geography textbooks. Many of these atlases were quite popular, and their wide distribution influenced the popular images.

There are usually one or two primary map publishers during any one period; Rand McNally and Hammond are today's. Since the works

[1] William H. Goetzmann, *Exploration and Empire* (New York: Knopf, 1966), p. ix.

[2] G. Malcolm Lewis, "The Recognition and Delimitation of the Northern Interior Grasslands during the Eighteenth Century," in *Images of the Plains: The Role of Human Nature in Settlement,* ed. Brian W. Blouet and Merlin P. Lawson (Lincoln: University of Nebraska Press, 1975), pp. 23–24.

of the primary publishers are most widely distributed, they have a considerable impact on the map- and atlas-buying public and indeed, their names become almost synonymous with maps. Thus, if the term "Great American Desert" appeared on an atlas distributed widely in schools, it was likely to have more authority than a designation in a minor publication.

American atlas maps of the nineteenth century have been criticized frequently for their inaccuracies, but the images presented by atlas maps are complicated. Atlas maps did not represent the view of a single explorer but, rather, were compiled from the recalled images of a number of individuals, both explorers and cartographers. While some of the data were taken from actual field observations, the maps were compiled from a variety of sources, usually including other maps. Subjective or even imaginative elements could be introduced in either recording or interpreting the data.

The distortions on nineteenth-century atlas maps were not necessarily introduced deliberately. Regardless of the honesty, objectivity, and accuracy of the explorers' reports, imagination modified the images. Those who translated the data to map form did not always make perfect translations or interpretations. As John Allen pointed out in "Lands of Myth, Waters of Wonder":

> Explorers may present a partial picture of an area by describing only what they have actually witnessed. Those who interpret their reports may have difficulty in adjusting the size of observed features to the scale of pre-exploratory *terrae incognitae*. Distortions may occur in translating what was seen on the ground (the landscape view) into an aerial perspective (the mapped view). What exploration makes known may expand in the imagination to encompass what remains unknown. Blank spaces are intolerable to the geographical imagination and people are tempted to fill them with imaginative extrapolation.[3]

Thus, a small area designated as treeless in an explorer's report could expand and become a large desert on a map.

Furthermore, the geography of a place is not entirely a result of

[3]John L. Allen, "Lands of Myth, Waters of Wonder: The Place of Imagination in the History of Geographical Exploration," in *Geographies of the Mind*, ed. David Lowenthal and Martyn Bowden (New York: Oxford University Press, 1976), p. 57.

what actually may be seen there but also of our *idea* of the place. "This geography of the mind can at times be the effective geography to which men adjust and thus, be more important than the supposedly real geography of the earth. . . . This is especially true of new environments; it is not what people actually see there so much as what they want to see or think they see which affects their reaction."[4] These mental images were a basis for decisions in many well-documented cases.

When we study the history of an area, too often we have before us a modern map created with precise instruments and methods, and we wonder at curious decisions which were made in the last century. If we observed the world in the same manner as those who made the curious decisions, its image reflected in the maps which they had before them—maps which were often distorted, sometimes inaccurate, frequently marked with legendary and fictitious places and features— we might improve our understanding of nineteenth-century thinking.

Although the atlases were erroneous in many respects (Bernard De Voto described them as horrendous), cartographers and publishers took some pains to be as accurate as possible and to advise the public of the problems inherent in the nature of maps. Mathew Carey, publisher of the first atlas produced in this country, stated in the preface of his *General Atlas* of 1817:

> My object is, to remove a general, though very inaccurate impression, respecting the execution of maps; and to prevent the purchasers of this work from judging it by too rigorous a standard—either of which would eventuate in disappointment, and excite unwarranted complaints. I desire to convey a clear and explicit idea of what I profess to furnish, that patronage be regulated accordingly; and that if I incur censure, it may not be for defects or imperfections, incident to and in fact, inevitable in every similar work, wherever published.

In the same vein, Benjamin Tanner, in an introduction to a series of map sheets issued for the *American Atlas*, stated that a map substitution had been made and the originally planned sheet would be issued with the next series because his information was not complete enough to publish.

[4]J. Wreford Watson, *Mental Images and Geographical Reality in the Settlement of North America*, Cust Foundation Lecture (Nottingham: University of Nottingham, 1967), p. 3.

Carl Wheat has criticized the early atlases for the apparent discrepancies from one map to another within the same atlas.[5] Two reasons may be given for these discrepancies. First, many of the early atlases featured maps of different areas compiled by different cartographers—a map of Texas might be drawn and compiled from one set of sources by one person, a map of North America drawn by a second cartographer using other source materials. Second, a common practice in the nineteenth century (as well as today) was using the same map in a number of different atlases in order to get more mileage from it. Thus it was possible in large atlases for maps showing roughly the same areas to disagree in some respects. For example, in the same atlas, parts of the Southwest might be found on maps of North America, the United States, Texas, and Mexico, and the maps could differ in some details. Such variations are a source of frustration to historians—and no doubt frustrated contemporary readers as well—but as the century advanced, atlases became more uniform in both form and content.

Changing Images

At the opening of the nineteenth century, Thomas Jefferson pressed for geographical mapping of the Trans-Mississippi West. In his library he had a copy of Aaron Arrowsmith's map of North America, published in 1801, in addition to other geographical works. Arrowsmith's map and the other maps of North America which appeared in atlases at that time showed an almost complete blank for the western portions of North America. Maps of the United States did not normally show land on the western side of the Mississippi River but instead ended abruptly. Maps of Mexico showed parts of modern Texas, New Mexico, and Arizona, and the Spanish settlements along the river valleys, but otherwise the southwestern quadrant was virtually unknown land. Those maps which did show some detail were largely conjectural.

The map in figure 4-1, by Philadelphia cartographer Samuel Lewis,

[5]For example, Carl I. Wheat, *Mapping the Transmississippi West: 1540–1861*, 6 vols. (San Francisco: Institute of Historical Cartography, 1957–   ).

Fig. 4-1. Map of western North America, by Samuel Lewis, in Aaron Arrowsmith, *New and Elegant General Atlas* (1804). *Courtesy Huntington Library, San Marino, California*

Aaron Arrowsmith's *New and Elegant General Atlas,* was published in London and Philadelphia in 1804. There is distortion of both size and shape, and since the map predates the expeditions of Pike and Lewis and Clark, much of the information in the West appears to be speculation. California is named New Albion, a place name which wanders on various maps.

Mathew Carey, a native of Dublin who settled in Philadelphia and founded a printing and publishing firm, is credited with publishing the first atlas to be produced in the United States (*American Atlas,* 1795). His firm continued publishing atlases until the late 1820s. Samuel Lewis drew many maps for Carey atlases, and figure 4-2 is from Carey's *General Atlas* of 1817. The information from Lewis and Clark's expedition had been published at this time, and the map appears to have had William Clark's map as a source. The probable boundaries of Louisiana are outlined; the Multomah River flows across the Great Basin. The Rocky Mountain system is shown branching, and another branching system appears in California. The detailed appearance of the map gives an impression of greater accuracy than actually exists.

By the 1820s atlases began to change in form; a more uniform appearance was evolving, with all maps on a consistent format. Although some maps still bore the name of individual engravers, as early as 1823 some atlases were produced under the supervision of a single person or firm, although this did not ensure consistency of maps. An example is Fielding Lucas, who had been a friend of Mathew Carey and drew many maps for Carey's atlases, produced his own *General Atlas* in 1823 (an act which jeopardized the friendship between the two men).

The maps in figures 4-3 and 4-4 represent the developments in Lucas's *General Atlas.* Figure 4-3 reflects much of the recent exploratory data, showing both "Highest Peak" and "James Peak" from maps of the Long Expedition. Highest Peak is now called Longs Peak, and James Peak, named for a Dr. James who ascended it on the Long Expedition, was called Highest Peak on maps of the Pike Expedition. Although the name James Peak appeared on maps for many years, that mountain is now known as Pikes Peak.[6] Lake Trimpanogos (Timpanogos), a legendary lake that appeared on early Spanish

[6]Carl I. Wheat, "Mapping of the American West: 1540–1857," *American Antiquarian Society Proceedings* 64 (April 1954): 19–194.

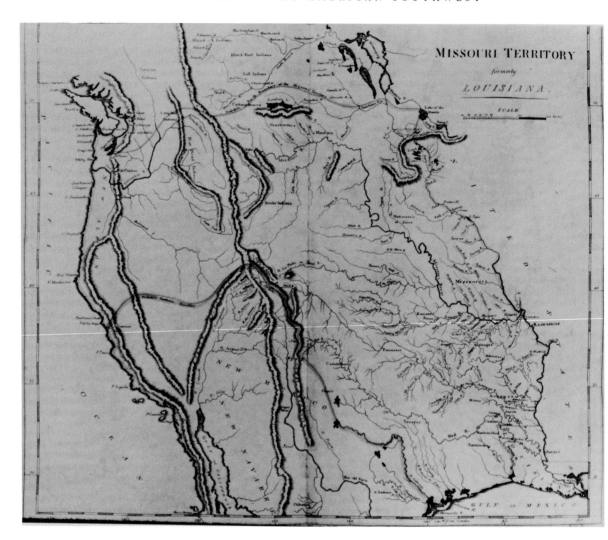

Fig. 4-2. Map of the Missouri Territory, by Samuel Lewis, in Mathew Carey, *General Atlas* (1817). *Courtesy Huntington Library, San Marino, California*

Fig. 4-3. Map of North America, from Fielding Lucas, *General Atlas* (1823). *Courtesy Huntington Library, San Marino, California*

Fig. 44. Map of the United States, in Fielding Lucas, *General Atlas* (1823). *Courtesy Huntington Library, San Marino, California*

maps, is shown but labeled "doubtful"; Robinson of the Pike Expedition adopted the name from Don Pedro Walker's map of 1810. The western limits of the Great Salt Lake are "unknown," while the Great Basin is called "Plains of Nuestra Señora and Luz," an interesting combination of Spanish and English.

Although figure 4-4 is also taken from Lucas's *General Atlas*, it is an example of the conflicting images which were found in early atlases. Lake Trimpanogos and the legendary river Buena Ventura are shown without disclaimers, and the Rocky Mountains differ in shape from the previous map. This is one of the earlier maps to show the Great Plains as the Great Desert—an image that persisted for three decades with some variations.[7] There has been some argument over the origin of the term; some have attributed it to Long, others to Pike, and still others to Dr. John H. Robinson of the Pike Expedition. Lucas also shows New Albion but applies the term only to Northern California.

Sidney Morse's *Atlas of the United States on an Improved Plan* was also published in 1823. Morse apparently based his atlas map (fig. 4-5) on John Melish's wall-sized map of the United States of 1816 (of which there were several variants). A connection between the River Buena Ventura and San Francisco Bay was described as "probably the connection between Atlantic and Pacific." The western limits of the Salt Lake were "unknown." This map also shows the location of Indian tribes, a practice which became common during the nineteenth century and probably confirmed in the reader's mind an image of the West and Southwest as a place heavily peopled by hostile Indians.

During the first half of the century, school atlases were popular. These ranged from slim volumes of less than a dozen maps to large works containing both general reference maps and thematic maps of climate, vegetation, animals, and the like. Normally these atlases were designed to accompany school geographies. William Channing Woodbridge's *School Atlas*, published to accompany his *System of Universal Geography*, went through several editions and had considerable influence in creating lasting images of the West. Figure 4-6 is from the

[7] Martyn J. Bowden, "The Great American Desert and the American Frontier, 1800–1822: Popular Images of the Plains," in *Anonymous Americans: Explorations in Nineteenth Century Social History* (Englewood Cliffs, N.J.: Prentice Hall, 1971), p. 71.

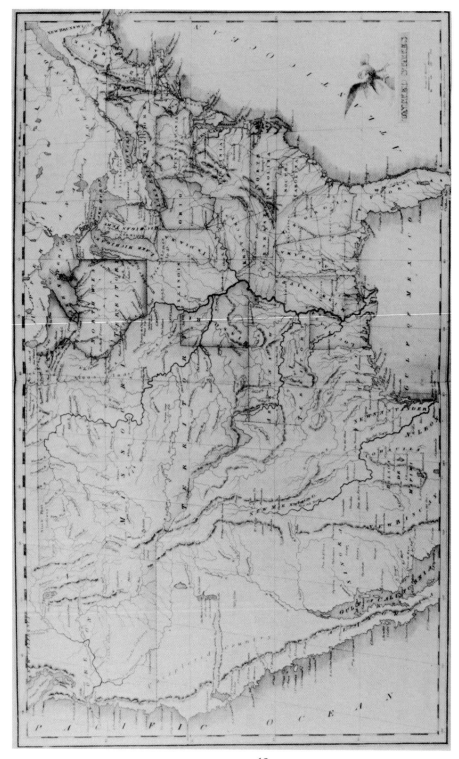

Fig. 4-5. Map of the United States, in Sidney Morse, *Atlas of the United States on an Improved Plan (1823). Courtesy Huntington Library, San Marino, California*

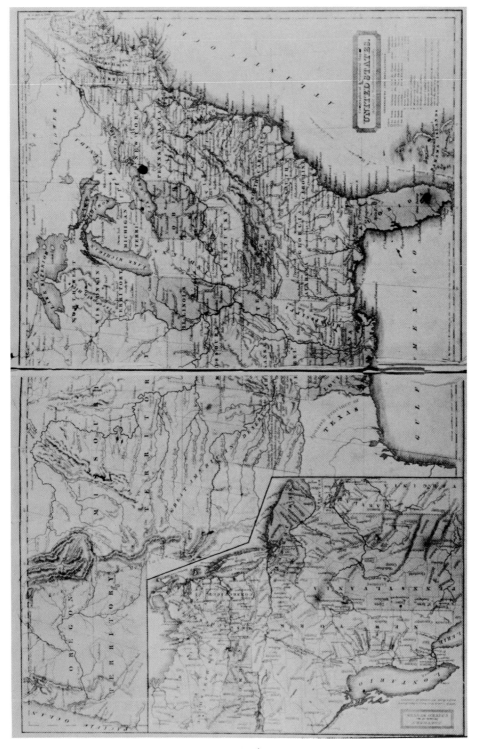

Fig. 4-6. Map of the United States, in William Channing Woodbridge, *School Atlas* (1833). *Courtesy Huntington Library, San Marino, California*

edition of 1833. The Great American Desert, covering nearly one-third of the area, is described as being "traversed by herds of buffalo and wild horses and inhabited by roving tribes of Indians."

One of the most respected commercial cartographers of the first half of the century was Henry S. Tanner. His *New American Atlas* of 1822, which was issued in sheets and included a geographical memoir describing his methods, is considered a landmark. Although Tanner showed some doubtful places on his maps, as did others, he included disclaimers indicating that the source materials were not of the same authentic character as the other materials used to construct the maps. Figure 4-7, taken from the later *New Universal Atlas* of 1836, is a considerably more authoritative work than the Woodbridge atlas in that there is less mythology and conjecture. The Great Basin is characterized as a "Sandy Plain." Indian tribes and districts are shown; a number of mountain peaks are indicated in the Rockies, and the Great Salt Lake is given the name "Lake Ashley."

Sidney Morse's *Cerographic Atlas* is important in American cartography for the printing method used—wax engraving which was to become so characteristic of American cartography for the next century. The map of Texas from this atlas (fig. 4-8) gives no indication of a desert; instead, areas in northern Texas are designated as "excellent land" or "rolling and fertile," a very different image from that presented by Woodbridge a decade earlier.

Figure 4-9, also taken from Morse's atlas but by a different author, shows various routes to the West as well as Indian tribes and the Paiuches Desert in the Great Basin. Lake Timpanogos, which died hard, reappears on this 1845 map.

By the 1840s the atlases of Samuel Augustus Mitchell—who apparently had acquired the rights to some of the Tanner plates—had become popular. Maps in the Mitchell atlases began to take on a more modern appearance. Figure 4-10 shows the Great Basin as the Great Interior Basin of California, and many Indian tribes are shown. In this *New Universal Atlas* of 1847, northwest Texas bears the inscription "this tract of country as far as the N. Canadian Fork was explored by Legrand in 1833, it is naturally fertile, well wooded and with a fair proportion of water." No Great American Desert here!

The firm of G. W. Colton shared Mitchell's position as primary commercial cartographic establishment until the 1880s. The 1866 sheet from Colton's *General Atlas* (fig. 4-11) shows southwestern Arizona as

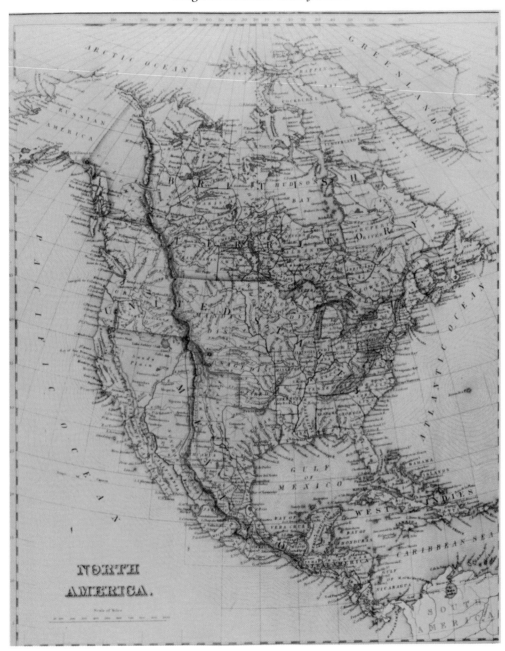

Fig. 4-7. Map of North America, in Henry Tanner, *New Universal Atlas* (1836). *Courtesy Huntington Library, San Marino, California*

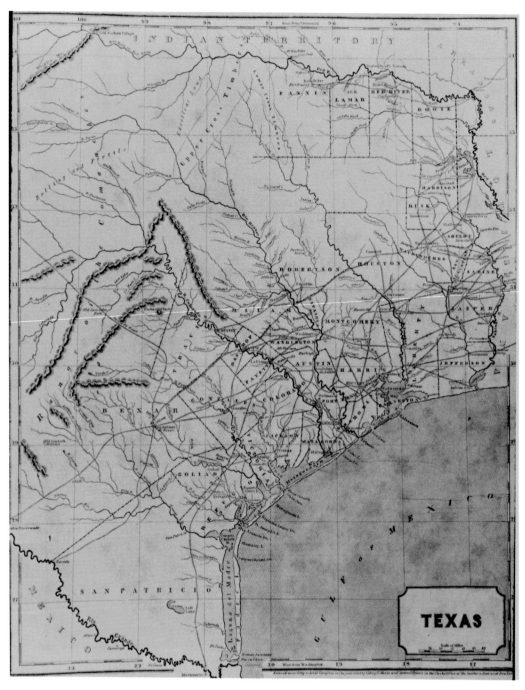

Fig. 4-8. Map of Texas, in Sidney Morse, *Cerographic Atlas* (1844). *Courtesy Hunt-ington Library, San Marino, California*

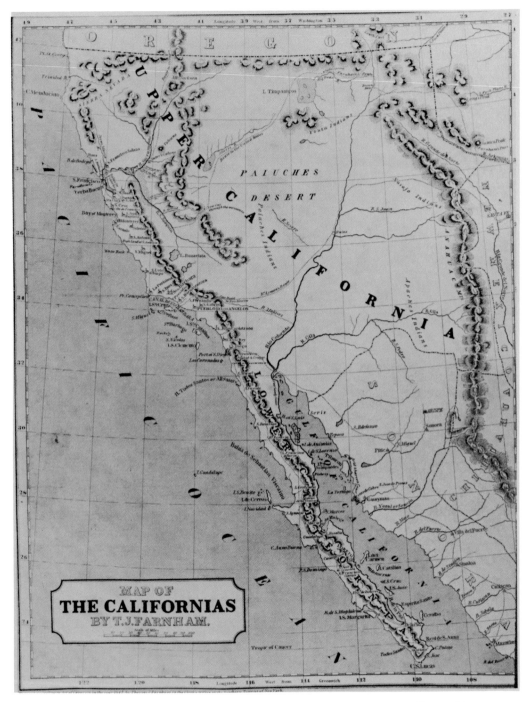

Fig. 4-9. Map of the Californias, by T. J. Farnham, in Sidney Morse, *Cerographic Atlas* (1845). *Courtesy Huntington Library, San Marino, California*

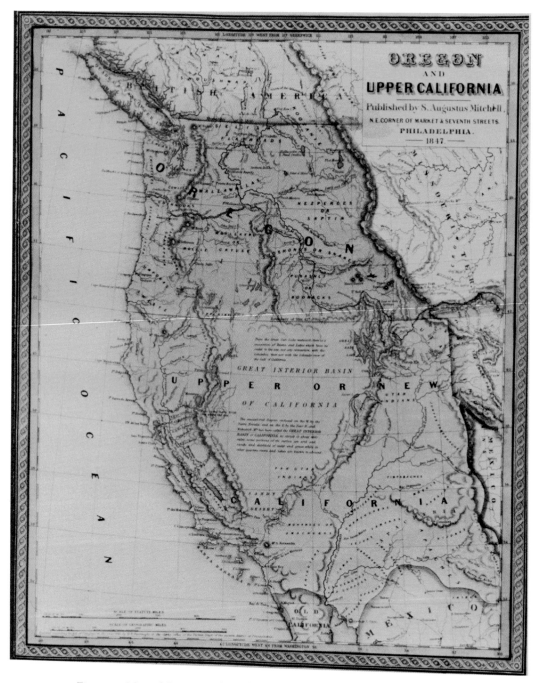

Fig. 4-10. Map of Oregon and northern California, in Samuel Augustus Mitchell, *New Universal Atlas* (1847). *Courtesy Huntington Library, San Marino, California*

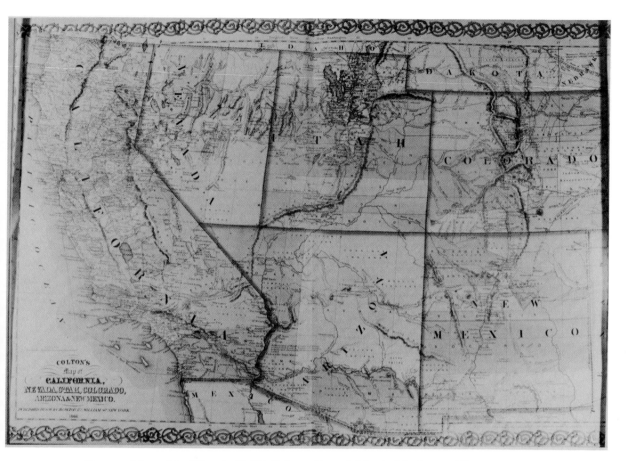

Fig. 4-11. Map of the West, in G. W. Colton, *General Atlas* (1866). *Courtesy Huntington Library, San Marino, California*

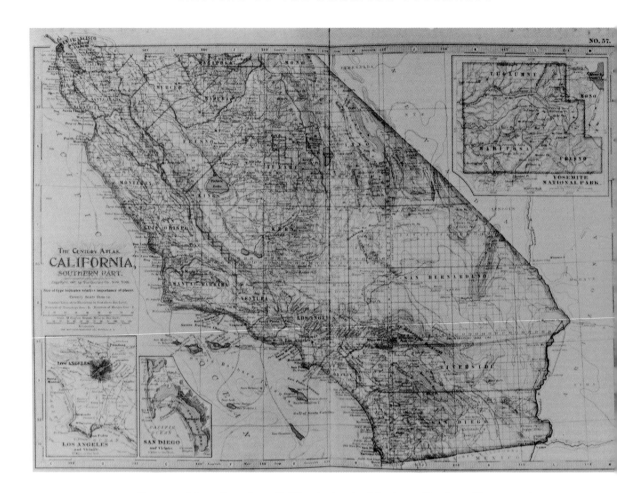

Fig. 4-12. Map of southern California, by the Matthews-Northrup Company, in *Century Atlas* (1897). *Courtesy Huntington Library, San Marino, California*

having abundant timber and water. Indian areas are shown, as was the common custom, and also places where people were killed by Indians —which undoubtedly created strong images of the area. In addition, the map indicates proposed railroad routes.

The last illustration (fig. 4-12) has a very modern appearance and rather closely coincides with our current images of California. Taken from the *Century Atlas* (1897) by the Matthews-Northrup Company, it was highly praised. Contour lines have replaced the more usual hachures, and the map indicates the township and range system as well as rail lines.

Within a hundred-year period, the image of the West and Southwest in American atlases had evolved from a vast *terra incognita* full of conjecture to one which closely resembled our own perceptions.[8] We, of course, assume our present images to be true, but perceptions are not static and even today our images are being altered by satellite photography, television, books, magazines, and atlas maps.

[8]Besides the sources already cited in this chapter, see John L. Allen "Exploration and the Creation of Geographical Images of the Great Plains: Comments on the Role of Subjectivity," in Blouet and Lawson, *Images of the Plains*; Martyn J. Bowden, "The Perception of the Western Interior of the United States 1800–1870: A Problem in Historical Geosophy," *Proceedings Assn. American Geographers* 1 (1969): 16–21; Bernard DeVoto, *Across the Wide Missouri* (Boston: Houghton Mifflin, 1947); Bernard DeVoto, *The Course of Empire* (Boston: Houghton Mifflin, 1952); Herman R. Friis, "The Role of the United States Topographical Engineers in Compiling a Cartographic Image of the Plains Region," in Blouet and Lawson, *Images of the Plains*, pp. 59–74; William H. Goetzmann, *Army Exploration in the American West 1803–1863* (New Haven: Yale University Press, 1959); Walter Ristow, "Early American Atlases," *Surveying and Mapping* (1962): 569–74; Walter Ristow, *Maps for an Emerging Nation* (Washington, D.C.: Library of Congress, 1977); Judith A. Tyner, "Trends in American Atlas Cartography," *Bulletin* Special Libraries Assn., Geography and Map Division, no. 125 (September 1981): 3–10; and J. Wreford Watson and Timothy O'Riordan, eds., *The American Environment: Perceptions and Policies* (New York: John Wiley and Sons, 1976).

Appendix
Selected Cartobibliography of the Works of Herman Moll Depicting the American Southwest

·

COMPILED BY DENNIS REINHARTZ

Works by Herman Moll

"America." London: 1697, 1700, and 1745.

Atlas Geographagus; or, a Compleat System of Geography, (Ancient and Modern) for America. . . . London: Eliz. Nutt, 1711–17.

"A Map of America According to ye Newest and Most Exact Observations . . ."

"A New Map of North America According to ye Newest Observations . . ."

"A Map of New France Containing Canada, Louisiana &c. in Nth. America . . ."

Atlas Manuale. . . . London: 1709, 1713, and 1723.

"The World in Planisphere . . ."

"America . . ."

"The Isle of California. New Mexico. Louisiane. The River Misisipi. and the Lakes of Canada. . . ."

"Mexico, or New Spain. Divided into the Audiance of Guadalayara, Mexico, and Guatamala. Florida. . . ."

Atlas Minor. . . . London: 1727, 1729, 1732, 1736, 1745, 1763?, and 1781.

"A New Map of the Whole World . . ."

"A New Map of ye North Parts of America according to the divisions thereof by the Articles of Peace in 1763 . . ."

"East & West Florida, and the North Part of the Gulf of Mexico, with adjacent Territories belonging to Great Britain & to Spain . . ."

"A Map of the West-Indies & c. Mexico or New Spain . . ."

The Complete Geographer; or, The Chorography and Topography of All the Known Parts of the Earth. . . . London: 1709 and 1723.

"The World in Planisphere . . ."

"America . . ."

"Isle of California. New Mexico. Louisiane. The River Misisipi. and the Lakes of Canada. . . ."

"Mexico or New Spain . . ."

"A Correct Globe. . . ." London: 1703 and 1710.

Dominia Anglorum in America Septentrionale specialibus mappis Londini primum a Mollio edita nunc recusa ab Homanianis Hered. . . ." London: 1729–90.

"A new map of the world according to Mercators projection, shewing the course of Capt. Cowley's voyage round it. . . ." London: 1699.

A System of Geography; or, a New & Accurate Description of the Earth in All Its Empire, Kingdoms and States. . . . London: T. Childe, 1701.

"The World in Plainsphere . . ."

"America . . ."

"The Isle of California. New Mexico. Louisiane. The River Misisipi. and the Lakes of Canada . . ."

"Mexico or New Spain"

Thesaurus geographicus. . . . London: 1695.

"The World in planisphere . . ."

"America . . ."

Thirty New and Accurate Maps of the Geography of the Ancients. . . . London: 1726, 1732, 1739, and 1755 and Dublin: 1736 and 1739.

"Orbis Tabula. . . ." Americas inset.

Twenty-Four New and Accurate Maps of the Several Parts of Europe. . . . London: J. Nicholson, n.d.

"A New Map of the World according to the New Observations . . ."

Untitled atlas of modern geography. London: 1700?

Untitled world map	"New Mexico"
"The World in Planisphere"	"Florida"
"America"	"New Spain"

A View of the Coasts Countries and Islands within the Limits of the South Sea Company. London: J. Morphew, 1711.

"A New & Exact Map of the Coast, Countries and Islands within ye Limits of ye South Sea Company . . ."

The World Described; or, A New and Correct Sett of Maps. . . .

"A New and Correct Map of the World . . ." [two hemispheres]. London: 1709, 1717?, 1720?, 1726, 1728, 1730?, 1732?, 1740?, 1754?, and Dublin: 1730 and 1741.

"A New and Correct Map of the Whole World Shewing ye Situation of its Principal Parts. . . ." London: 1720?, 1726, 1728, 1730?, 1732?, 1740?, 1754?, and Dublin: 1730 and 1741.

"To the Right Honourable John Lord Somers Baron of Evesham in ye

County of Worcester President of Her Majestys most Honourable
Privy Council &c. This Map of North America According to ye
Newest and most Exact Observations is most Humbly Dedicated
by your Lordship's Most Humble Servant Herman Moll Geographer. . . ." London: 1712, 1717?, 1720?, 1726, 1728, 1730?, 1732?, 1740?,
1754?, and Dublin: 1730 and 1741.

"A New Exact Map of the Dominions of the King of Great Britain on
ye Continent of North America. . . ." "Beaver map." Americas inset. London: 1715, 1717?, 1720?, 1726?, 1730?, 1732?, 1740?, 1754?, and
Dublin: 1730 and 1741.

"A Map of the North Parts of America claimed by France under ye
names of Louisiana, Mississippi, Canada and New France with ye
Adjoining Territories of England and Spain. . . ." London: 1720?,
1726, 1730?, 1732?, 1740?, 1754?,and Dublin: 1730 and 1741.

"A Map of the West Indies or the Islands of America in the North
Sea. . . ." London: 1717?, 1720?, 1726, 1730?, 1732?, 1740?, 1754?, and
Dublin: 1730 and 1741.

"A New & Exact Map of the Coast, Countries and Islands within ye
limits of ye South Sea Company. . . ." London: 1711?, 1717?, 1720?,
1726, 1730?, 1732?, 1740?, 1754?, and Dublin: 1730 and 1741.

"Nine Plates of Charts of the Principal Sea-Coasts throughout the
whole Earth. . . ." London: 1740? and 1754? Unknown?

Maps by Moll in the Works of Others

Binger, Herman. *The Louisiana Purchase and Our Title West of the Rocky
Mountains, with a Review of Annexation by the United States.* 2 vols. Wash-
ington, D.C.: Government Printing Office, 1900.
"Part of a Map by Herman Moll, English Geographer, published in
London about the year 1710."

Bowen, Emanuel. *A Complete System of Geography.* . . . 2 vols. London: Innys,
1747.
The maps are essentially the same as and based on those in the 1723
edition of Moll's *The Complete Geographer.* . . .

Burpee, Lawrence J. *The Search for the Western Sea.* . . . New York: Macmil-
lan, 1936.
"A New Map of North America . . ."

Dampier, William. *Collection of Voyages.* . . . London: James and John Knapton, 1729.
"A Map of the World. . . ." Funnell map of 1707.

"A Map of the Middle Parts of America. . . ." Funnell map of 1707.
"America . . ."
"A View of the General & Coasting Trade-Winds in the Great South
   Ocean. . . ." Early thematic map.
————. *A New Voyage Round the World.* . . . London: 1698, 1703, and 1729.
"A Map of the World Shewing the Course of Mr. Dampier's Voyage
   Round It: From 1679, to 1691 . . ."
"A Map of the Middle Parts of America . . ."
Defoe, Daniel. *The Life and Strange Surprising Adventures of Robinson Crusoe,
   of York, Mariner.* 4th ed. London: W. Taylor, 1719.
"A Map of the World, on wch is Delineated the Voyages of Robinson
   Crusoe"
Dodd, William E. *The Old South: Struggles for Democracy.* New York: Mac-
   millan, 1937.
"A map of the West Indies &c. Mexico or New Spain. . . ."
Funnell, William. *A Voyage Round the World.* . . . London: James Knapton,
   1707.
"A Map of the World . . ."
"A Map of the Middle Parts of America . . ."
"America . . ."
Harris, John. *Navigantium atque itinirantium bibliotheca; or, A Compleat Col-
   lection of Voyages and Travels.* . . . London: Thomas Benet, 1705.
"A New Map of the World According to Wrights alias Mercator's Pro-
   jection &c. . . ."
"A New General Chart for the West Indies . . ."
Moore, Jonas. *A New Systeme of Mathematiks Containing . . . A New Geog-
   raphy . . . With Maps to Each Country and Tables of Their Longitude and
   Latitude.* . . . London: 1681.
"America . . ."
Oldmixon, John. *The British Empire in America.* . . . 2 vols. London: 1708,
   1741, and 1744.
"A New Map of North America . . ."
Osborne, Thomas. *A Collection of Voyages and Travels.* . . . London: 1745.
"A New Map of the World according to the Newest Observations . . ."
"A General Chart of the Sea Coast of Europe, Africa & America . . ."
"A Chart of ye West-Indies or the Islands of America in the North Sea
   &c . . ."
"A View of ye General & Coasting Trade-Winds, Monsoons or ye Shift-
   ing Trade Winds through ye World, Variations &c. . . ." Early the-
   matic map.
"A New Map of North America According to the Newest Observa-
   tions . . ."

"A Map of Mexico or New Spain Florida now-called Louisiana and Part of California &c . . ."

"A Map of New France Containing Canada, Louisiana &c. in Nth. America . . ."

Rogers, Woodes. *A Cruising Voyage Round the World.* . . . London: 1712 and 1718.

"A Map of the World with the Ships Duke & Duchess Tract Round it. from 1708 to 1711. . . ."

Salmon, Thomas. *Modern History.* . . . 5 vols. London: 1725–39, and Dublin: 1727–39 and 1755.

"A New Map of the Whole World . . ."

"A View of the General Trade-Winds, Monsoons or Shifting-Winds & Coasting-Winds through ye World, Variations &c . . ."

"America . . ."

"A Map of the West-Indies &c. Mexico or New Spain. . . ."

"A Map of the World. . . ." Dublin: 1755.

"Florida. . . ." Dublin: 1755.

Wright, Thomas. *The Famous Voyage of Sir Francis Drake.* . . . London: H. Slater, 1742.

"A New Map of the World according to the New Observations . . ."

The Mapping of the American Southwest

was composed into type on a Compugraphic digital phototypesetter in twelve point Janson with two points of spacing between the lines. Janson Italic was selected for display. The book was designed by Jim Billingsley, typeset by Metricomp, Inc., printed offset by Thomson-Shore, Inc., and bound by John H. Dekker & Sons. The paper on which the book is printed bears acid-free characteristics for an effective life of at least three hundred years.

TEXAS A&M UNIVERSITY PRESS
College Station